Amanda Emiliana Santos Baratelli

LAND, STATE AND CAPITAL

Amanda Emiliana Santos Baratelli

LAND, STATE AND CAPITAL

THE CENTRALITY OF LAND RENT IN ECONOMIC AND POWER RELATIONS IN THE MUNICIPALITY OF TRÊS LAGOAS/MS

ScienciaScripts

Imprint

Any brand names and product names mentioned in this book are subject to trademark, brand or patent protection and are trademarks or registered trademarks of their respective holders. The use of brand names, product names, common names, trade names, product descriptions etc. even without a particular marking in this work is in no way to be construed to mean that such names may be regarded as unrestricted in respect of trademark and brand protection legislation and could thus be used by anyone.

Cover image: www.ingimage.com

This book is a translation from the original published under ISBN 978-3-330-99520-8.

Publisher:
Sciencia Scripts
is a trademark of
Dodo Books Indian Ocean Ltd. and OmniScriptum S.R.L publishing group

120 High Road, East Finchley, London, N2 9ED, United Kingdom
Str. Armeneasca 28/1, office 1, Chisinau MD-2012, Republic of Moldova, Europe
Printed at: see last page
ISBN: 978-620-3-81554-2

I dedicate this work to Geography. The science I fell in love with and found myself in.
I dedicate the fruit of this achievement to my family, for whom I fight every day.

ACKNOWLEDGMENTS

I'm writing this thank-you note as a letter for you to read in a few years' time. Therefore, I intend to make it as explicit as possible how I feel at this point in my career and, above all, to mark out who were/are the people who walked side by side with me at this stage.

I've been studying Geography since 2016 and it's only now that I feel safe to measure the importance of this encounter in my life. That's right, an encounter. Geography has been my great encounter so far. It's where I trained and built myself. It was through Geography that almost all the answers came and, even better, the questions too. I continue to walk my path in the place where my heart beats, my eyes admire and my feet tread. The geography I believe in is done on my feet, in practice, or rather, in praxis.

I'd like to point out that knowing Geography only as a science didn't touch me deeply. It was the relationships played by it and through it that made me fall in love. Therefore, it is the people who share Geography with me that I have to thank. As with geographical concepts, what good is space if it's not inhabited? What good is form if there is no society to give it meaning? What good is geography if we don't do it every day?

I've had the pleasure of sharing this journey with some very important people. So I'd like to start by thanking my family, who formed me socially and enabled me to broaden the vision I have today. I'd like to start with my mother, Márcia, who has always respected the importance of my studies and supported me throughout this journey. To my father, Marco, whom I have no words to describe, I share the struggle to make this dream come true. I dedicate this achievement to my sister, Leticia, the initial forerunner of my relationship with my studies, in whom I have always looked up to. I would also like to thank my sister-in-law, Winnie, for her support on this journey.

Like my family, I am very grateful to my teachers, who enabled me to learn everything I know today. Ever since I was a child, I have admired and respected my teachers, and now, some time later, I know that I made the right choice to mirror and consolidate myself in a profession made with love and respect. I would especially like to thank the following professors: Professor Sedeval Nardoque, with whom I had the joy of writing an article and who helped me, with great attention and care, towards the end of this work; Professor Marta Inez, for her important contributions to my qualification; Patrícia Milani, for accompanying me in my undergraduate research and for her attention throughout my career; Thiago Santos, for making me fall in love with Latin American studies; Mauro Soares, Marine, Jodenir Calixto, Rafaela, Fred, André, Patrícia and Vitor. In particular, Professor Edima, who kindly provided me with data on her research in partnership with SEBRAE, the report of which had not yet been published.

Last but not least, I would like to thank Professor Rose, someone I have always looked up to and admired for the strength that goes hand in hand with sensitivity. Thank you for teaching me Agrarian Geography and the virtues of an educator. Your support at different times was essential to my various achievements.

To my research and geography friends, I would especially like to thank Dener and Luiz, who have been by my side at all times. Sharing the dream and the realization of the master's degree with you was very special. I would also like to thank my research companions Jhiovanna, Diego, Letícia, Danilo, Luciene and Mieceslau. All of you were fundamental to the realization of this research, whether through your support in words and/or in the battle to acquire data that is still very nebulous in Brazil.

I would also like to thank my friends from Andradina and Três Lagoas, who have been present in my daily life during this pandemic, celebrating my achievements and listening to me talk non-stop about my research lol. Respectively, Layne, Luana, Gabi, Flávio and Maju, Amanda, Gabriela and Karina. I'll toast with you.

Finally, I would like to thank the Federal University of Mato Grosso do Sul for being part of my story. This year I end my career here, but I will always remember UFMS. I would like to thank the Coordination for the Improvement of Higher Education Personnel (Coordenação de Aperfeiçoamento de Pessoal de Nível Superior) for allowing me to carry out this research with a grant. On behalf of the coordinator, Professor Patrícia Mirandola, I would like to thank the Postgraduate Program in Geography (PPGGeo UFMS/CPTL).

To all the faculty and students of the UFMS CPTL Geography course, I share a little piece of this achievement. In this Geography, from the countryside, I became a professional, politicized myself, especially at the Collective Management Meetings of the Association of Brazilian Geographers (AGB) and consolidated myself as a researcher. I will be eternally grateful!

I end my words and this work with the feeling of having done my duty, but knowing that the struggle for land, for social rights and for justice will always be present in my actions. May we free society from the bonds of this savage capitalist system. May we succeed in decommodifying the land and making it a place where life can be reproduced.

SUMMARY

When analyzing the history of Brazilian territorial formation, we find evidence that shows the constant participation of rentier agrarian oligarchies in political spheres since the mid-20th century, seeking decisions that favored their oligarchic interests. In order to understand how these oligarchies came to control the state, it was necessary to understand the dynamics of the consolidation of the patrimonialist state, whose core relationships are based on political patronage, through favors, nepotism and private interests (MARTINS, 1994; FAORO, 1958). The patrimonial characteristic of the Brazilian state is no exception to the rule of what the state is as an institution of power. According to authors such as Pachukanis (2017), Harvey (2005) and Poulantzas (1985), the state machine is in constant dispute between the social classes. However, the power of the hegemonic classes, such as landowners and industrial capitalists, has dominated control of the state, since, in a class alliance, they have used it to secure benefits. Pachukanis (2017) also points out that the constitution of the rule of law as the power of the rule of law serves as an ideological haze to hide and execute the interests of the dominant classes that make up and command the state. It is worth pointing out that the state corrupted by patrimonial relations has a unique characteristic in Brazil, since the classic dispute between antagonistic classes, such as industrial capitalists and landowners, has become a pact in Brazil: the land-capital alliance (MARTINS, 1994). The fact is that this dispute between land and capital, represented by antagonistic classes, has not ceased; however, these classes are confused in their interests, supported by public power, since the state enables and encourages them to play a dual role in the machinery of capitalism, either as landowners and/or industrial capitalists, thus increasing the appropriation of property for the private realization of class reproduction. In addition, the state subsidizes the actions of these classes, avoiding direct conflict between them, without completely extinguishing the conflict inherent in the class struggle. The aim of this paper is to investigate the realization of the land-capital alliance in the municipality of Três Lagoas, as well as its particularities in terms of the dynamics of earning land rent and profit. Considering the expansion of the eucalyptus-cellulose business in the municipality, we sought to understand this process in terms of the relationship between the countryside and the city, understanding that these two fractions are dialectically complementary in the single capitalist territory. In order to consolidate the proposed objective, theoretical and methodological procedures were adopted to understand the territorial formation of Três Lagoas, whose heritage goes back to the process of division of the state of Mato Grosso and the formation of Mato Grosso do Sul. On a local scale, a group of families with social distinction and representative of economic and political power were defined: the Thomé family, the Salomão family and the Prata Tibery family. These families belong to the local elite and play crucial roles not only in rentier capitalism, but also in urban business, particularly after the expansion of eucalyptus and cellulose in Três Lagoas. At first, the historical consolidation and social prestige of these families was analyzed, highlighting how their enterprises exercise double accumulation, combining the rentier function with profit extraction. This movement of accumulation is possibly more related to class unity, rather than class alliance (MARTINS, 1994). In this sense, it was also necessary to understand the presence of the territorialization of capital in the territory of Tres-Lago, materialized in the installation of two pulp and paper agro-industries, Suzano Papel e Celulose and Eldorado Brasil, respectively. The conclusion is that the territorialization of industrial capital has not changed the dynamics of local economic and political power. Landowners have managed to diversify their businesses, especially in urban areas, in order to maintain and increase the possibility of earning income for the private realization of their class reproduction. Thus, the wealth derived from the territorialization of the pulp-paper complex continues to be amalgamated with the interests of traditional groups in the municipality, while part of the population continues to be dispossessed of land and live with the social ills produced by/in the "national pulp capital".

Keywords: Land-capital alliance; Agrarian oligarchy; Patrimonial state; Land concentration; Rentier capitalism; Territorialization of monopolies.

SUMMARY

INTRODUCTION

The work in question is part of a research project related to Agrarian Geography, the center of the debate of which is based on the assumption of the existence and permanence of an agrarian question in Brazil. In order to carry out this theoretical exercise, it was necessary to adopt a research method, which in this case was the historical-dialectical materialist method. The method established is part of the theoretical foundations laid down by Karl Marx, whose basis was Hegelian dialectics.

The use of the historical-dialectical materialist method presupposes the understanding and use of fundamental concepts such as totality, contradiction, reproduction, hegemony and singularity.

The dimension of totality established by Marx (2008) is based on the synthesis of multiple determinations, instituted in the performance of society's economic structure. However, for Marx, the notion of totality is made up of heterogeneous parts, resulting from the unequal and contradictory movement of capitalist relations of production. It is therefore worth pointing out that contradiction is not the materialization of duality, but the realization of the unique movement of society in its relations of production, through singularity and universality.

The category of reproduction is based on the execution of social processes that are the basis of society. In this sense, as not all social relations are based on capitalist premises, there is also the reproduction of non-capitalist relations. In this case, we can highlight the relationship played out by the peasantry, whose central social process is the non-capitalist relationship with the land, in other words, the use of the land as an instrument for the reproduction of life (MARTINS, 1981). There are other dynamics reproduced in society that also demarcate the contradiction as a product of society, such as the expansion of latifundia to earn land income, which represents a contradiction in the expanded reproduction of capital. However, in the case of Brazil, landowners have also managed to reproduce capitalist practices, above all through relations with the territorialization of capital (OLIVEIRA, 2007).

The consolidation and maintenance of the capitalist system was achieved through hegemony, which turned the system in question into a dominant force of thought that conditions the social classes (GRAMSCI, 1991). In this case, the presence of hegemony can be seen in the actions carried out by the rule of law, which drives the interests of the ruling class, supported by the law, and blocks other social interests.

The singularity of reality, despite being limited to the social phenomenon being studied, does not mean that the universality of the processes is dismissed. It is not disconnected from the totality; on the contrary, the formation of the totality is based on multiform processes and particular manifestations. This is therefore the task of researchers: to unveil how universality materializes in singular representations:

> The researcher's task is to unveil how universality is expressed and concretized in singularity, or, more than that, how universality is expressed and concretized in the diversity of singular expressions (PASQUALINI, 2015, p.364).

In this sense, the aim of this work is to draw on concepts from the historical-dialectical materialist method to achieve the following objective: to investigate the presence of the land-capital class alliance in the municipality of Três Lagoas, as well as its particularities in terms of the dynamics of earning land rent and profit, considering the recent expansion of capitalist activities related to the eucalyptus-cellulose complex. It is therefore proposed to understand this process through the relationship between the countryside and the city, understanding that these two fractions are dialectically complementary in the single capitalist territory.

The municipality of Três Lagoas, located on map 1, had its territorial formation based on extensive cattle ranching and the concentration of predominantly low-productivity land. In the last decade, the dynamics of livestock farming have been combined with new dynamics in the countryside, which have also had an impact on the city. The changes in the countryside came about as a result of the territorialization of the pulp and paper complex, respectively the companies Suzano Papel e Celulose and Eldorado Brasil.

Map 1 - Três Lagoas (MS): location of the municipality and its headquarters

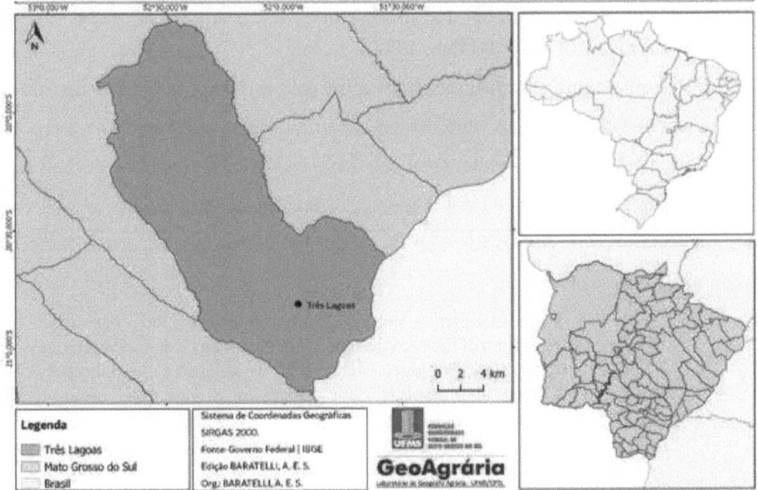

However, even with the territorialization of the eucalyptus-cellulose monopoly, local power - economic, political and ideological - has remained with traditional families, whose main activities are related to capitalist land ownership.

In Brazil, according to Martins (1989), land represents the center of economic relations, because the expanded reproduction of capital is based on the realization of land income. In this way, the land-capital alliance was consolidated in the country as the backbone of the economy and of power, allowing industrial capitalists and oligarchic groups to live together, a relationship mediated by state subsidies.

> (...) Brazil's economic modernization was conditioned by the interests of the national and international bourgeoisie and the landowners who ended up determining a pattern of dependent capitalism, which further tied conservative interests to the logic of "[...] permissiveness with speculative movements, patrimonialist obliquity [...]." (LESSA; DAIN, 1998, p. 260 apud PIRES, RAMOS, 2019, p. 418).

In addition to the assiduous presence of the state in mediating relations between industrial capitalists and landowners, in Brazil, this dynamic has unique characteristics, since the Brazilian Bourgeois Revolution was initiated by landowners, as evidenced by Fernandes (2020). In this way, industrial capitalists and landowners merged, exercising both economic activities, in other words, they united the sources of land income and profit (MARTINS, 1994).

It's worth pointing out that the land-capital pact is only possible because of the characteristics of the Brazilian state, which is patrimonialist in nature. The premises of this state model stem from clientelist relations, based on personal favoritism that becomes political (FAORO, 1958). Thus, according to the author, the groups that represent public power govern for their own benefits and maintain the pact between classes, mitigating conflicts.

> Patrimonialism is pulverized into an isolated localism, which the retraction of the secular state accentuates, in such a way as to convert the public agent into a client, within an extensive clientelist network. The colonel uses his public powers for private ends, often mixing the state organization and its treasury with his own assets (FAORO, 1958, p. 757).

It is important to emphasize that capitalist dynamics impose changes on the patrimonialist order based on clientelism and state control, so it is not a question of continuity or mere reproduction of the oligarchic past. It is in this sense, that we must understand the conservative modernization of Brazilian agriculture and the land-capital pact. In other words, we need to recognize that the agrarian elites are modernizing their relationship with the capitalist market, but without changing the historical basis of the authoritarian land monopoly.

> (...) the modernization of large-scale farming has been portrayed as "[...] the implementation of the Prussian way in Brazil [which] has been called the 'conservative modernization' of Brazilian agriculture by various authors." (RAMOS, 2007, p. 35). According to Ramos (2007), the national agricultural pattern followed the path of the modernization of large agricultural property without any fragmentation of the national land structure. (PIRES, RAMOS, 2019, p. 418).

To assume that the theoretical debate on the land-capital pact (MARTINS, 1981) is at the heart of the causes of the concentration of wealth and social exclusion of the majority of Brazilians, is to consider that the state is structurally capitalist and maintains the privileges of the powerful. However, it is not a question of considering that this is because we are facing a historical inheritance (the so-called "vice of origin") materialized in the permanence of the anachronistic, corrupt state, an obstacle to the development of the market. On the contrary, Martins teaches in "O poder do atraso"

(The power of backwardness), 1994, that the modern and the archaic are part of the dialectical movement that runs through the elites and the state, combining elements of continuity and discontinuity in a tragic way for society. Therefore, it is the permanence and constant updating of the class pact around the control of land rent and profit that has allowed the rich to remain rich even in the face of changes and arrangements in the country's territorial dynamics.

This means that the entry of paper company capital into the Bolsão territory, particularly in Três Lagoas, did not make the power of the landowners unfeasible; on the contrary, it reaffirmed the pact through the ideology of progress and development. The possibility of accommodating the class interests of landowners and capitalists - which in essence are disparate - created an economic environment favorable to the transformation of the former into capitalists, especially in the city, which became big business. This unity would not have occurred so quickly without the active participation of the state through public funds (financing, exemptions, donations, etc.) - this is the modern face of patrimonialism and also the basis for the conversion of unproductive estates into productive ones. It is in this sense, that of the state as financier and entrepreneur, that the large public loans to the eucalyptus and cellulose agribusiness in Três Lagoas should be interpreted.

In order to provide theoretical support for the proposed debate, this dissertation has been divided into three chapters. The first, entitled "The rentier nature of Brazilian capitalism and the agrarian question in Mato Grosso do Sul", reviews the theory of rentier capitalism, i.e. the role of land as the center of the capital production process and the state as its support.

In the second chapter, we sought to understand the process of dividing the state of Mato Grosso do Sul based on the continuity of the practices carried out in what was still Mato Grosso. It was therefore a question of investigating, above all, the land structure and the actions of the state's leaders in keeping it concentrated.

The third chapter sought to explain how the municipality of Três Lagoas has followed the tradition of land concentration since the formation of MS, and today has a land structure that symbolizes the monopoly of land. And historically, it highlighted the presence of economic and political power in the control of the state as a representation of private class interests, to the detriment of the rest of society.

In order to highlight the proposition of the third chapter, namely wealth based on capitalist land ownership (rentier capitalism) and urban business, it was necessary to identify the profile of traditional and influential families in the municipality of Três Lagoas. In this way, three families were selected who are part of the local elite and who own businesses related to the dynamics of earning income from the land and also in the practice of making a profit: the Thomé family, the Salomão family and the Prata Tibery family. Although these families are unique, there are important similarities, since they have characteristics linked to the landowning class, as well as currently expanding their business in the capitalist market to also appropriate profit via the land-capital alliance/unit. However, despite this process essentially maintaining the strategies defined theoretically by Martins (1994), in the case of Três Lagoas there is a particularity, since the alliance is, in fact, the class unit. In other words, in Três Lagoas, it's not necessarily the urban industrial capitalists who become landowners, but the local landowning class itself which, having a monopoly on land, takes advantage of the contemporary capitalist dynamic of expansion in the cellulose sector to diversify its sources of wealth, especially in the urban area, combining income and profit. Consequently, the territorialization of the eucalyptus-pulp complex did not represent a change in the essence of local power, much less a reduction in traditional territorial dominance[1] . What was achieved was a strengthening of the land-capital unity as a source of class accumulation.

From a methodological point of view, the following procedures were adopted: a) investigating the families' rural property holdings; b) acquiring official documents from the Três Lagoas District Land Registry Office; c) drawing up cartographic maps for selected areas of these families; d) conducting interviews with people who have knowledge of the local situation .[2]

In addition to these procedures, it was also necessary to gather data on urban areas, since the landowning families' enterprises are not limited to the countryside. In this way, we also tried to relate the presence of their capitalist interests in urban areas. We therefore sought to understand how these groups of landowners relate to the

---

[1] This was the main reason for the difficulty faced by pulp companies in acquiring land in Três Lagoas at the start of the eucalyptus plantation expansion process - a difficulty materialized in the increase in the price of land. See Kudlavicz, 2011.

[2] Four interviews were conducted, with a local lawyer, a former civil servant from Três Lagoas and two residents of neighborhoods in the city.

dynamics of the expansion of eucalyptus monocultures and the territorialization of the cellulose-paper complex.

With regard to the process of acquiring data and information for this research, it is important to reflect on the difficulties faced in acquiring information on surnames considered "traditional" in the city. There is an ideological pact in the municipality of Três Lagoas (and I fear in many other municipalities) which translates into a strangeness when seeking information about traditional families who own land, especially about their possessions and assets. Even in the public sector, whose principle should be transparency, information is given in a nebulous way, with a certain amount of fear, with various questions: "for what?" and "for whom?".

This research was carried out between the years 2020-2022, i.e. it was produced during the period in which the world was (and still is) in a tragically pandemic scenario, so the difficulties were multiplied, especially in finding research subjects willing to corroborate. In short, the task of constructing a social survey in a country in isolation was arduous and, above all, locating subjects who knew the city and, what's more, were aware of the importance and need for access to public information in order to advance scientific knowledge. Within this list of obstacles, the quest to conduct interviews, so that data and information could be better analyzed, was the greatest challenge of this work. For example, in order to obtain information and reports about rentier practices in Três Lagoas, it was necessary to assure the interviewee that the conversations would not be recorded, a situation that made it impossible to analyze and use in the work. This corroborates our view that the practice of using land as a store of value - the operational center of rentier capitalism - is a debate that is still obscure in society, typical of the authoritarianism that accompanies it.

In this sense, I ask the following question: what is research for? Or rather, who is research for? Research helps us get to know society. It serves to document facts and reflections that will serve the evolution of knowledge. It serves to create and update concepts. It serves to compose new theories in order to understand the material and immaterial elements of society. In this way, I think it serves everyone who believes in the need to democratize land, agrarian reform and social justice as a way of overcoming rentier capitalism. However, ideological constraints do not allow society to understand this. Fortunately, I was able to count on the courageous people who believe in the importance of research and knowledge for the advancement of society in favor of the common good. That's right, courage, because research requires courage.

CHAPTER 1: THE RENTIER NATURE OF BRAZILIAN CAPITALISM AND THE AGRARIAN QUESTION IN MATO GROSSO DO SUL

The aim of this chapter is to highlight the rentier nature of Brazilian capitalism as a basis for the appropriation of land income. It also aims to understand the dynamics of land speculation and land concentration built into the formation of the Brazilian state.

It is therefore necessary to understand the capitalist state as an institution of power taken over by the interests of the ruling classes. For this, some authors were essential, such as Harvey (2005), Pachukanis (2017) and Poulantzas (1985).

The patrimonial characteristic of the Brazilian state has allowed the concentration of land and the maintenance of inequalities in relation to its access, as revealed by the works of Martins (1981, 1994; 2020), Faoro (1958), Leal (1997), Pires, Ramos (2009), Motta (2015) and Castilho (2012).

Historically, the condition of a patrimonial state has been based on the use of the public machine to favor the private interests of the ruling elite. This is how clientelist relations come about, based on personal favoritism that also becomes political favoritism (MARTINS, 1994).

The formation of the Brazilian state is mainly marked by the participation of agrarian oligarchies who, at times, grew their estates with the use of legislation and public money. The presence of agrarian oligarchies, or their appointees, in public power reveals the rentier side of power relations in Brazil - with land concentrated, they get rich by extracting income from the land (CASTILHO, 2012).

In order to provide an opportunity to debate the concepts related to understanding land rent, we sought support from classic readings on the subject, such as Marx (2017), Oliveira (2007), Martins (1981; 1989; 2020).

Land rent is the social tax paid by society as a whole. However, in order to be able to earn rent, there needs to be inequality between those who own land and those who are dispossessed. At this point, the state serves to deny the poor the right to land, whether for housing and/or for the reproduction of life (OLIVEIRA, 2007).

In Brazilian capitalism, land has been converted into a commodity, capable of generating income and, through the territorialization of agro-industrial enterprises, also generating profit. It is in this sense that Brazilian capitalism presents itself as a paradox, since it inverted the classic logic of the capitalist system and began to establish itself on the basis of rentier capitalism, whose land remains at the center of social processes (MARTINS, 1989).

The possibility of using land to despoil nature allows for the generation of income and profit. Profit is generated through the reproduction of capital and the production of commodities. In this way, we also have to analyse the occurrence of this process in the municipality of Três Lagoas, particularly in the analysis of the territorialization of the pulp and paper complex, which was born subsidized by voluminous resources from the state, be it federal, state or municipal.

1.1 THE process of capital production based on the monopoly of land: agrarian oligarchies and land concentration

For Poulantzas (1985), although the state represents an institution of control, it is not a static and immutable "thing". It is a movement of social relations, a product of society, generated by the relationship, including disputes between antagonistic classes. In this way, the state emerges as the ideology of an institution responsible for moderating the conflict between antagonistic classes, between the bourgeoisie and the proletariat, and goes beyond the condition of being "neutral" in order to maintain the common interests of both classes.

Although the state is a social formation, representative of both classes, its control has been taken over by the bourgeoisie. According to Pachukanis (2017), the premise that the state is the result of a dispute between classes does not answer all the questions about what the state actually is. For the author, the dispute between antagonistic classes - the bourgeoisie and the proletariat - resulted in the victory of the ruling class. They are the ones in control of the state. The state has not established its balance between the class struggle by becoming a super-class force, quite the opposite, it has been transformed into an instrument of the bourgeoisie, since "it is advantageous to create an ideological haze and hide behind the screen of the state its class domination." (PACHUKANIS, 2017, p.179).

Still from the perspective of Pachukanis (2017), the use of the state as a "neutral" tool of social control allows for the constitution of the rule of law, whose power lies in the empire of the laws, thus: "The state machine is in fact realized as impersonal "general will", as "power of law" etc. to the extent that society represents a market." (p.182).

The Laws, which should represent the "general will", actually serve as an instrument for defending the interests of the bourgeoisie. However, the social conception that justice is neutral, as, for example, in the representation of the scales of Themis - goddess of justice, who appears with her eyes closed and with a scale representing equality between both sides - distorts the true interest in which the rule of law is used.

> he legal state is a mirage, but a mirage that is totally convenient for the bourgeoisie, because it replaces the decaying religious ideology and hides the rule of the bourgeoisie from the masses. The ideology of the legal state is more convenient than the religious one, because not only does it not reflect the totality of objective reality, but it also relies on it. Authority as the "general will", as the "force of law", insofar as it is realized in bourgeois society, represents a market (PACHUKANIS, 2017, p. 185).

In practice, therefore, the state has become a regulator and, at times, a promoter and guardian of capitalist relations of production, creating mechanisms, such as laws, to maintain the civil order that guarantees the maintenance of the capitalist system (HARVEY, 2005). In this sense, the capitalist state model has always remained submissive to the ruling classes:

> The state that originates from the need to keep class antagonisms under control, but which also originates in the midst of the struggle between classes, is usually the state of the economically ruling class, which, through its resources, also becomes the politically ruling class, and thus obtains new means of controlling and exploiting the oppressed classes. The ancient state was first and foremost the state of the slave masters to control the slaves, just as the feudal state was the organ of the nobility to oppress the peasant serfs, and the modern representative state is the instrument for exploiting wage labor for capital. (ENGELS *apud* HARVEY, 2005, p. 78).

The state has become an instrument of domination for the ruling class, which uses it to prioritize its private interests over the collective ones. Harvey (2005) explains that in order to consolidate state domination, two strategies are created. The first is the expression of institutional autonomy and the condition of authority, in which even its officials are above civil society. The second strategy is to turn the interests of the dominant ruling class into an ideology of collective interests, thus highlighting the imposition of particular interests and domination between classes (HARVEY, 2005).

In order to maintain the capitalist order, based on exchange value, the state establishes legal mechanisms that regulate the ownership of private property, as in the case of land, which in the capitalist system allows a common good - use value - to be transformed and sold as a commodity - exchange value. The rule of law, using its legal apparatus, creates social control mechanisms and constitutes benefits, based on the law, to meet the demands of capitalist market interests.

In addition to the legal mechanisms that regulate private property, the state also creates ideologies about the condition of equality between subjects from different social classes, such as the use of the terms "natural person" and "legal person", which serve to characterize citizens who are equally free to carry out economic exchanges. Although the terms are based on principles of equality, the exchanges made between worker and employer are unbalanced from an economic point of view. The worker sells their labor in exchange for a salary, which serves as a subsidy to maintain their minimum living standards, such as food, housing, etc., while the owner of the means of production, through unpaid work, extracts surplus value from the worker, which is converted into profit and returned as capital to be invested and generate more profit/capital for the employer. Thus, the relations of equality in the employment contract serve to maintain the balance between antagonistic classes and preach a false sense of freedom (HARVEY, 2005; MARTINS, 1981). Thus:

> The capitalist state must necessarily support and enforce a legal system that encompasses concepts of property, the individual, equality, freedom and law, corresponding to the social relations of exchange under capitalism (HARVEY, 2005, p. 81).

The legal and ideological regulations determined by the capitalist state serve to maintain the balance between the classes, making it appear that their relations are based on social contracts of equality, for example, when workers feel free to sell their labor. In other words, they don't feel exploited, because they understand that they have been given the power to choose.

With the action of state regulations, the masses feel controlled by the legal, political, ideological and, not least, military apparatus. Military power is a strong ally of the capitalist type of state, according to Poulantzas (1985), since at times of systemic hegemonic crises and fierce division between the classes, the capitalist type of state breaks down. Still according to Poulantzas (1985), the transition to regaining control of the state goes through two stages. At first, the state of exception, in which control of political processes is concentrated in the state, even suspending the electoral process and the rule of law. The dismantling of democratic institutions and the exacerbated use of violence by the military serve as a way of mystifying the fragile condition of the state in crisis.

With the process of regaining control With the process of regaining control through configurations of the capitalist type of state, this state goes through yet another transitional phase, called by Poulantzas (1985) authoritarian statism, whose characteristics are the concentration of activities in the executive branch, the decline of the rule of law, the fall of social representations based on political parties and the growth of parallel forces that cut across the structure of the formal organization. In addition, in order to guarantee time for the agreements and alliances within the state's representative "clans" to be reconciled, military power is used to control the masses and their demands.

Although at times of crisis of hegemony and splits between antagonistic classes, social demands arise from the people, the state, when it re-establishes its capitalist state order, tends to act with the aim of controlling and re-establishing the common social order, and for this it creates public policies based on the state of exception. The state of exception in question, as studied by Oliveira (2004), is different from the state of exception of Poulantzas (1985). For Poulantzas, the state of exception arose at times of crisis and to maintain its hegemony through violence. The state of exception, cited by Oliveira (2004), is a permanent characteristic of a state that is subjugated to the ruling class and international institutions of economic control and power, such as the IMF (International Monetary Fund) and the World Bank. In this case, the state's actions are limited, materialized in public policies to contain (and maintain) poverty, without being able to generate major social transformations, since investments must be concentrated on the development of capitalism.

In this sense, it is important to understand the state as a movement, a product of social relations and the historical context in which it is inserted (POULANTZAS, 1985). However, one cannot overlook the fact that the state tends to be controlled by the dominant classes, who use it as an ideological instrument to guide their individual interests, making it an important cog in the wheel of the capitalist system.

Class domination is present in the capitalist system and in state control beyond the official domination of the state machine. Domination is present in the relations of dependence of governments on banks and capitalist groups, in the dependence between worker and employer and, not least, in the effective composition of the state apparatus being closely linked to the ruling class (PACHUKANIS, 2017).

According to Faoro (1958), the effective presence of members of the ruling class in the composition of the state and the relationships of personal subordination, in which social rights are suppressed for the sake of personal favours, characterize the type of state whose ties are patriarchal.

Patriarchal relations of dependence gave rise to the political model of a patrimonial state, where the center of relations is based on the will and interests of the head of state and, in this way, favors, personal favors, nepotism, etc. prevail and "the absence of a distinction between the ruler and the state (private *versus* public) prevails in patriarchalism". (SARAIVA, 2019, p. 339). Thus:

> The political community conducts, commands and supervises business, as its own private business at first, and as public business later, along lines that are gradually demarcated. The subject, society, is understood within the framework of an apparatus to be exploited, manipulated and, in extreme cases, sheared. From this reality is projected, in a natural flourishing, the form of power, institutionalized in a type of domination: patrimonialism, whose legitimacy is based on traditionalism - this is how it is because it has always been (FAORO, 1958, p.866).

In the case of Brazil, the state has a patrimonial character, inherited from the economic and social relations carried out by the king in Portugal. The use of Brazil as a Portuguese colony - "the king's business" - was integrated into the patrimonial structure. As an example of this patriarchal relationship, the king transferred the land grants in the hereditary captaincies to his noblemen, based on a relationship of "favoritism" (SARAIVA, 2019). As Martins (1994) corroborates:

> The king's representatives granted the land in the name of the king's interests and not in the name of the people's needs. There is no sesmaria request from the colonial period in which the applicant does not justify it with services already rendered to the crown, in the war against the Indian, in the conquest of the territory (p. 23).

However, despite the granting of land in the hereditary captaincies, according to Fonseca (2019), they were not configured as a domain, only destined for use, which lasted for a short time, between the years 1534 and 1549, without leaving important marks on the Brazilian land structure.

In contrast to the hereditary captaincies, the division of the sesmarial system determined structural conditions in the Brazilian agrarian question. The division of sesmarias was the first institutionalization of Brazilian land and lasted in the country for several years (1530-1822). The model of granting land began in Portugal in 1375, but with different objectives to those used in Brazil. In Portugal, land was granted for a fixed period of time and with the obligation to cultivate it. In Brazil, however, even with the obligation to cultivate, the main purpose of the distribution of sesmarias was based on the conquest and invasion of indigenous territories - systematically violated by Portuguese colonization (MARÉS, 2003).

Like this:

> The use of sesmarias, therefore, was Portugal's way of promoting the conquest of Brazilian territory. [...]. The land was granted so that the beneficiary could come to Brazil to occupy it, in the name of the Crown, producing export goods on a large scale, even if it meant persecuting, enslaving or killing indigenous populations, and generating African slavery and hunger. (MARÉS, 2003, p. 59).

The constitutionalities established by the sesmarias were, from the outset, disrespected in Brazil, since in addition to the concessions for use of the Crown's monopoly areas, several donations of immense areas were made, determining the establishment of private properties, even within the sesmarial regime - lands of the king (MOTTA, 2012).

According to Marés (2003):

> Therefore, in the 16th century there were already signs that the granting of sesmarias could create problems in the organization of the country's land, but the limitation imposed on the concessions, which had to be the size of the beneficiary's capacity to take advantage of the land, was useful [...]. The concessions continued to disobey this criterion and in the 17th and 18th centuries they ended up becoming a source of latifundia creation. If in the beginning they served as an instrument of external conquest, being used for Portugal to seize territory, once Portuguese power was established they became an instrument of internal conquest, **serving to consolidate the power of the latifundia because the concessions became a distribution from the elite to themselves, as an exercise of power and its maintenance.** (MARÉS, 2003, p. 62. My emphasis).

The patrimonial system remained present throughout the construction of the Brazilian state. With the favored relationships granted by the king to the grantees, they began to establish autonomy and dominate the land and regional power. In order to increase their dominance and succession, they began to arrange marriages between their children, thus merging important surnames to take control of the chambers and judicial decisions. Furthermore, the process of growth of the estates and domination of power was mediated by corruption and processes of extreme violence, yet the state continued to favor its favorites, without interference (SARAIVA, 2019).

During the process of land grabbing and the illegal granting of internal possessions to the elites, other land occupations were also taking place. Rural workers - who because of their social class were not entitled to access land under the sesmarial regime - began to occupy idle land in order to reproduce family life, producing food and selling surpluses. These workers were called posseiros and occupied the idle land between the coffee monocultures. With the increase in squatters occupying the land, the state began to regulate the situation, encouraging the acquisition and control of possession, which led to various conflicts between squatters and landowners (MOTTA, 2012).

The various conflicts and the territorial size of Brazil made it impossible to control the land and oversee the sesmarias, which is why various illegal actions continued to take place. However, with the country's independence in 1822, the sesmarias regime came to an end, as Portuguese laws were no longer in force in Brazil (MOTTA, 2012).

In this period, with the absence of legal regulations on land ownership; the abolition of slavery; the arrival of immigrants in Brazil and the need for the state to become more developed from a legal point of view, in 1843 the Land Law project was drafted and presented (FONSECA, 2019).

The Land Law, enacted in 1850, was the genesis of capitalist land ownership in Brazil, serving to make land a commodity and institutionalize the huge estates distributed and constituted (by illegal practices) under the sesmarias regime, through land grabbing practices[3] . In addition, the regulation of land prevented former slaves, peasants and immigrants from occupying it, unless they bought it. The regulation of private property, mediated by exchange value, determined that no one could seize another's property by force, i.e. it could only be bought - exchange value.

---

[3] The word "grilagem" has a curious origin. In order to age documents, fraudsters put them in a box (or drawer) filled with crickets. The action of the insect makes the papers look older than they are (CASTILHO, 2012, p. 60).

In addition, with the end of the sesmarias regime and the enactment of the Land Law of 1850, the state instituted that "vacant land could not be occupied by any title other than purchase" (MARTINS, 2021, p. 44). The wastelands therefore returned to the monopoly of the state, which was already controlled by the agrarian oligarchies (MOTTA, 2012).

The absence of legal regulations that institutionalized land use until 1850 meant that the sesmeiros increased their land holdings over the years, until they constituted immense latifundia, forming the class of landowners in Brazil (FONSECA, 2019). In order to prove ownership of the land grabbed, which occurred in 1854 at the parish registry, landowners falsified property titles, with dates prior to the parish registry and with documents recognized by official notaries, made possible by the industry of falsifying property titles. For peasants, squatters and former slaves, the vast majority of whom were illiterate, legal strategies were not possible, mainly due to financial limitations and the lack of knowledge that was denied to them, causing their land to become vacant (MARTINS, 2021).

During the period when slavery was in force in Brazil, between 1535 and 1888, landowners measured their power by the number of slaves they owned. In this way, the income of these owners was capitalized in the purchase of slaves - merchandise - and investments for the slave trade. The exploitation of unpaid labor conferred profit on the coffee and sugar cane plantations. Thus, "this is effectively capitalized income, a capitalist form of income, income that takes the form of profit" (MARTINS, 2021, p. 34).

Not only did slaves generate profit through unpaid labor, but they also served as mortgage collateral for farmers to obtain investments from banks to expand their coffee farms. In this way, the number of slaves guaranteed credit to landowners. However, the slave regime was shaken by the prohibition of the slave trade through the Eusébio de Queiroz Law in 1850. The cessation of the slave trade caused slave prices to rise disproportionately, so landowners immobilized part of their capital in the purchase of captives in order to secure bank loans. According to

> The rise in the price of slaves increased the basis for obtaining mortgage loans, while the expansion of coffee loans depended on a greater immobilization of capital, in the form of income capitalized in the person of the captive. This situation, therefore, did not benefit the landowner, but the trafficker [...] (MARTINS, 2021, p. 43).

The contradiction of the need to immobilize more capital in the purchase of slaves in order to guarantee bank loans for the expansion of coffee plantations represented a systematic contradiction for the plantation owners, since this process benefited them. Therefore, the alternative to ending the contradiction in question was the abolition of slavery. This was a decision made by the landowners in conjunction with representatives of the state (MARTINS, 2021).

With the end of slavery in Brazil in 1888, via the signing of the Golden Law by Princess Isabel, landowners were no longer able to capitalize on the captive's income in order to grant bank loans.

It is worth noting that the abolitionist strategy was not aimed at freeing blacks from the bonds of the system of labour exploitation in Brazil, since the former slaves were marginalized in capitalism's new labour model. For the capitalist mode of production, together with the practices of the rule of law, the liberation of slavery was a strategy to make workers believe in the ideology that they were free to sell their labor power, in order to legitimize exploitation by capital (MARTINS, 2021).

From this perspective, the new working model of capitalist relations needed a worker who saw work as a virtue of freedom. However, a society whose center was slave relations was based on the discourse that the foundations were based on master and slave and, therefore, there would be no way to make this type of worker viable in Brazil (MARTINS, 2021). It was with this discourse that the process of immigration of foreign workers to the coffee plantations began, between 1886 and 1914, known as the settlement labor regime.

This strategy of bringing in immigrant workers, as was the case in the settlement system, maintained the basis of the exploitation of labor that existed in Brazil, but without it being slave labor, but also without it being able to be classified as wage labor, because the earnings of these workers could be monetary and non-monetary, according to Martins (2021):

> It cannot be defined as a system of wage labor, since wages in cash are, in the capitalist process of production, the only form of remuneration for labor power. This is because the colonato was characterized, as will be seen in detail later, by the combination of three elements: a fixed payment in cash for the coffee plantation, a proportional payment in cash for the quantity of coffee harvested and the direct production of food, as a means of living and as a marketable surplus by the worker himself, therefore, a pre-capitalist peasant component in the employment relationship. (p. 35-36).

The impossibility of mobilizing capital to buy slaves and the impossibility of doing unpaid work absolutely - through slave labor - meant that the landowners increased their percentage of profit by increasing the productivity of their crops. As a result, what mattered to them was to increase their plantation areas by opening up new farms, a job done by the settlers.

The expansion of coffee farms through the work of settlers set the precedent for land to become an alternative for immobilizing income to use it as a mortgage and access bank loans, because land had become a commodity (MARTINS, 2021).

Although land began to be used as a mortgage and an alternative for capitalizing on income during the settlement system and the expansion of coffee plantations, from the very beginning of capitalist relations, land was conditioned to function as a commodity, even though it was not an object of value.

For Marx (2017, p.676) "land ownership is based on the monopoly of certain people over defined portions of the globe with exclusive spheres of their private will [...]". In this way, even though land is not the product of labor, it plays the role of a private commodity, subject to use through purchase and/or lease.

Land is the basis of capitalist production, since products are extracted from it - through cultivation - and buildings are built on it. Therefore, in order for the land to be used for production purposes, the capitalist must pay the taxes imposed by the landowner for the use of his monopoly. Thus:

> This lessee-capitalist pays the landowner, the owner of the land he exploits, a sum of money fixed by contract (in exactly the same way that the borrower of money capital pays for it at a fixed interest rate) in exchange for being allowed to invest his capital in that particular field of production. This sum of money is called land rent, regardless of whether it is paid for arable land, building land, mines, fishing grounds, forests, etc. It is paid for all the time during which the landowner has lent, rented by contract, the land to the tenant. In this case, ground rent is the form in which land ownership is economically realized, the form in which it appreciates in value (MARX, 2017, p. 679).

In this sense, land is a commodity marked by contradiction in typically capitalist relations, since it is an obstacle to the expanded reproduction of capital, because in order for the capitalist to buy or lease a piece of land, he needs to immobilize part of his capital in capitalized land rent (MARTINS, 1981).

In this way, the capitalist who buys the land also becomes the landowner and guarantees the extraction of taxes from the property, materialized in the form of land rent. In this sense, those who own land privately have the right to earn land rent, which is the tax paid by society as a whole, extracted through social surplus value, based on the unequal relationship between those who own land and those who don't (MARTINS, 2021).

Therefore:

> In capitalism, land, which is also transformed into a commodity, has a price, but no value, because it is not a product created by human labor. Capitalist ownership of land is capitalized income; it is the right to seize an income, which is a fraction of the social surplus value and therefore payment subtracted from society in general. This is because there is a class that privately owns the land and only allows it to be used as a means of production (whether rented or not) by levying a tax: the capitalist land rent. That's why, under capitalism, buying land is buying rent in advance. When we are faced with land grabbing, this process reveals its true character: the 'free' way of access to rent, access to the anticipated right to obtain payment of rent, without even having paid to be able to earn it. In the same way, but in the opposite direction, possession is the act of those who don't want to pay the rent or don't accept the condition that they have to pay it in order to produce (OLIVEIRA, 2007, p. 66).

The rentier nature of capitalism is referred to by Marx as "parasitic capitalism", since land is used as a "store of value", according to Oliveira (2001). Rentierism is one of the ways in which capital is produced through non-capitalist relations, made possible by private land ownership and land concentration. Land is therefore used as a commodity, but it cannot be assigned value, since value is socially necessary labor and there is no labor that produces land. In this way, land is given a price (MARTINS, 1981).

The price paid by the tenant will be determined by the qualities and characteristics of the portion of land traded. In this case, to define the form of land rent earned, Marx (2017) differentiated land rent into absolute rent, differential rent I, differential rent II and monopoly rent.

The absolute rent of the land is part of the process in which the landowner intends to earn the land income all at once, by selling it. When calculating the sale of land, the landowner adds to the price the income capacity it could generate. According to

In conclusion, it can be said that absolute land rent comes from the contradictory interests between the classes or fractions of classes in capitalist society and the monopoly power of one of them, exercised in the agricultural production process over the soil. It can be earned, as we have already seen, by putting the land to work, or it can be earned all at once by selling it. This is because in the capitalist mode of production land, although it has no value (as it is not the product of human labor), does have a price, and buying it gives the owner the right to charge society in general for the income it can provide. In other words, when you buy land, you buy the right to earn income from the land (OLIVEIRA, 2007. p. 57).

Differential rent I is related to factors that determine the highest price, especially with regard to agricultural production. For this type of rent, the fertility of the soil and the location of the plot add to the land rent. Thus, land with a higher fertility index and better locations fetches higher prices, while land without these differential attributes fetches lower prices. Land rent keeps the possibility of earning the highest price or not always in an unequal relationship with other plots (OLIVEIRA, 2007).

Differential income II is related to the amount of investment in the land that will need to be made by the tenant and/or landowner. In this case, the use of capital to improve the land, through the use of agricultural inputs, soil correction, improved management, etc., will determine the percentage of income earned. The landowner who owns the land without the need for capital investment will reduce expenses and increase the percentage of income earned (OLIVEIRA, 2007).

Finally, for monopoly land rents to be earned, the land needs to have unique and exceptional characteristics that are difficult to find elsewhere. These characteristics increase the price of the land exponentially, because it is considered a "special" commodity in capitalism. Oliveira (2007) uses the example of the production of Port wine, which is only produced in the Porto region of Portugal, under specific soil and climate conditions, to illustrate how the monopoly price of the land and the commodity produced is determined by a single factor (OLIVEIRA, 2007).

In addition to the forms of land income determined by Marx, the centrality of the private appropriation of land in capitalist relations of production highlights other ways of generating income from the appropriation of territory. According to Oliveira (2007), in addition to earning income from the monopoly of captive land, the process of subjection to income can also take place in two ways, namely the monopolization of the territory, in which the income from peasant labour is extracted in the commercialization of commodities (agricultural products) and the territorialization of the monopoly, which occurs through the installation of large agricultural enterprises and predatory extraction.

The coffee business under the settlement system materialized both ways of earning income, both through the sale of the fruit and through the predatory practice of intensive agriculture, with the expansion of monocultures. In addition, the owners of the coffee plantations realized that the increase in the search for access to land among workers would lead to a rise in its price, making it possible to increase bank credit for capital costs on coffee plantations, resulting from mortgaged land. It was therefore established that for immigrants to become owners of small plots of land, they would need to earn it through their work (MARTINS, 2021).

Thus, for Martins (2021, p. 47) "if the land was free, labor had to be a slave; if labor was free, the land had to be a slave. The captivity of the land is the structural and historical matrix of the society we are today". The conversion of land into mortgages to guarantee credits for the expansion of coffee plantations and, consequently, with the increase in the price of plots, allowed the landowning class to intensify the exploitation of territorial income. Therefore:

> In Brazil, capitalized territorial rent is not essentially a transfigured feudal inheritance. It was engendered in the midst of the crisis of slave labor, as a means of guaranteeing the subjection of labor to capital, as a substitute for the expropriation of the peasantry, which, with the advent of capitalism, created the mass of disinherited people able to enter the labor market of the new society. Here, property had the function of forcing the creation of a supply of free and cheap labor for large-scale farming. Here it was the substitute for primitive accumulation in the production of labor power, with the same function: the expansion of capitalism would only be possible with the emergence of a mass of free workers because they were free of the means of production to work on their own, subject, therefore, to the need to work for capital in order to survive (MARTINS, 2021, p. 48).

With the replacement of the slave as capitalized income for the captivity of the land, landowners invested in the expansion of their farms, since now they didn't have power over the person of the worker, but over the product of their work. According to Martins (2021, p. 59), "[...] more important than ownership over the worker was securing the work that creates wealth, that creates value". Therefore, the opening of new farms and the gain in capital through the amount of coffee harvested, for example, materialized the free worker's ability to make a profit through surplus value. Its market value, therefore, belongs to the fruit produced/harvested (MARTINS, 2021). Thus:

> Territorial income arises from the metamorphosis of capitalized income in the person of the slave; it therefore arises as a form of tax capital from commerce and not from the trafficker, as the acquisition of the right to exploit labour, as opposed to the right of ownership over the person of the worker. Slave ownership is configured as ownership of the land as a means to extort labor from the worker and not to extort income from him in the form of labor and product. Capitalized territorial income is not an instrument of leisure, but an instrument of business. The best evidence of this can be found in the fact that the landowner who lives by renting out his property to capitalist tenants is still a relatively rare phenomenon in Brazilian society, which is widespread in the more characteristically entrepreneurial sectors of the agricultural economy. Furthermore, the landowner and the capitalist coexist (MARTINS, 2021, p. 48).

The understanding that the generation of capital should occur through an increase in coffee productivity, and therefore the expansion of new farms, increased the demand for labor. The recent past of slavery and the investments made by landowners to bring in settlers to work on the plantations brought violence to the fore. Farmers transformed the investments to bring in workers into personal debts owed by the settlers, so that in order for the worker to move to another farm, he would have to be bought by another farmer who would pay off his debts (MARTINS, 2021).

Due to the problems of the servile relationship imposed by the landowners on the settlers, which threatened the civil liberty of the workers, and the growing demand for labor, the state began to intervene. Thus, it began to subsidize the arrival and housing conditions of these workers, as shown by Martins (2021):

> The immigrants, however, were preferably settled in official colonies, on smallholdings. The government paid for transportation costs from Brazil to the place where the immigrant and his family would settle. As well as paying for and financing the land and the initial costs of maintaining the family, it maintained a regime of guardianship over the settler, usually for a period of two years (p. 57).

In this way, state intervention in the training of the coffee workforce consisted of diverting public resources to a single sector of the economy, located mainly in the Southeast of the country. The inequality of the state's investment in the other regions caused heated discussions, and the official colonies were sometimes far from the planting areas, making it difficult to use the available labor (MARTINS, 2021).

Farmers who cultivated sugar cane plantations in the Northeast began to demand the labor of these workers for their region, and with this, the state changed the policy for distributing labor in the country. Private colonies began to be set up inside the plantations, according to the landowners' requests to the immigration sector. Thus:

> Farmers would no longer have to bear the costs of immigration, which would be subsidized by the government, leaving them free to immobilize their capital in the person of the settler, in the form of capitalized income, with expenditures on agents, shipping companies etc. Instead of sending immigrants recruited by government agents to official colonies, they began to be sent to the coffee plantations themselves, at the request of the farmers in charge of immigration (MARTINS, 2021, p. 58).

The labor subsidy offered by the state allowed landowners to convert the amount that would be capitalized for the arrival of workers into the expansion of coffee farms. The settlers were sent to work on opening up new farms, dealing with the land and planting new coffee plantations. The creation of new farms stimulated the land market in the 1880s, with illegal practices such as land grabbing on the rise in the 1870s. The fact is that the landowners, who at first defended the servile relationship between worker and employer, in a traditional view of landowners, began to introduce into their characteristics the capitalist mentality, whose central valuation lies in the appropriation of labor for surplus value (MARTINS, 2021).

> Only with the intervention of the state was it possible to break the captive labour circuit, socializing the costs of training the workforce and creating the conditions for the establishment of free labour and the labour market. The intervention of the state in the training of the workforce for the coffee plantations in fact represented the provision of subsidies for the formation of capital for the coffee enterprise (MARTINS, 2021, p. 59).

Therefore, one cannot discuss captive land and mortgage guarantees without considering the transition from captive worker to free worker, made possible by the participation of the state.

In the case of Brazil, although land was already a guarantee of income, the expansion of coffee plantations and the increase in the price of plots intensified landowners' search for land, so that the amount of land replaced the guarantee of power that had previously been determined by the number of slaves. In capitalism, the exploitation of workers generates profit, since the surplus value of their labor is appropriated by the owner of the means of production. Land, on the other hand, generates land rent, the product of the social surplus value earned in the unequal relationship of access to land since the dawn of the capitalist system. Capitalism is therefore not just a mode of commodity production, it is a mode of production based on the exploitation and realization of surplus value, be it individual or social (MARTINS, 1981).

The fact is that in the dynamics of earning rent from the land, the state is an important agent in making the actions of capitalists and/or landowners viable. In the act of territorializing the monopoly, the state acts as a partner of capital, from a financial point of view with subsidies and from a legal point of view with flexible legislation, such as the installation of large companies in the countryside, as in the case of Suzano S/A and Eldorado Brasil, in the municipality of Três Lagoas (KUDLAVICZ, 2014).

Thus, the use of public funds for capitalists and landowners occurs because Brazil has a state dominated by the hegemonic classes in order to maintain their wealth.

The history of clientelism and the establishment of the landowning class consolidated parasitic relationships in Brazilian politics, such as the rentier agrarian oligarchies. The oligarchies came to dominate regional political relations, hold public office, influence votes, etc. The old colonels held (and still hold) financial power and public influence, making the interests of their own group the collective interest, thus maintaining the continued reproduction of the oligarchies' hegemony in the institutions of public administration. Thus:

> It is a group whose genesis is marked by hybridity: on the one hand, it appears to be just a sector educated and specialized in the functions of administration and politics that stands out in the nucleus of families of large rural landowners, where its social origin lies; on the other hand, its strengthening is closely linked to the patrimonial use of the state, the control of appointments to public office and the traffic it enjoys in the upper echelons of the national administration (REIS, 2007, p. 51).

The rentier agrarian oligarchies, structured since the mid-19th century, grew their wealth through the privilege of active participation in the constitution of governments, which meant that the laws favored them directly, or the infractions they committed were not penalized. In this sense, even with the Proclamation of Independence (1889) and, almost fifty years later, with the emergence of a New State (1937-1945), the old relationships inherited from Portugal were maintained. The Brazilian state, therefore, was born and remained in a patriarchal cradle.

The agrarian oligarchies gained a foothold in Brazil. At first, they dominated the municipalities, influencing the electoral process and promoting their nominees. With the favor of the governors, they expanded their interference to regional politics, and thus consolidated a national territory to guide their particular interests. The growth and contamination of the political system was such that it was only possible to win an election with the support of the traditional oligarchs (MARTINS, 1994; SARAIVA, 2019).

Thus, since the origins of political relations in Brazil, favoritism and privileges for the ruling classes have been prominent. According to Martins (1994), favoritism was mediated by political patronage, which "[...] has always been and is, above all, preferably a relationship of exchanging political favors for economic benefits" (p. 29). Clientelist relations materialize ties between the powerful and the rich.

In clientelism, the rich, even when they are not elected, become part of government decisions, above all, to guide their interests through actions in the interest of private wealth, some of which can be classified as corrupt. Therefore:

> The tradition of a political system based on the confused relationship between public and private wealth has been the basis on which this relationship has given way to procedures that are beginning to be classified as corrupt (MARTINS, 1994, p. 40).

Furthermore, the patrimonial base and the strong presence of agrarian oligarchies became a permanent condition of the Brazilian state. From the Empire to the Republic. At the beginning of the Empire, under Dom Pedro I (1822-1831), there was talk of institutional restructuring, but patrimonial relations were sovereign, preventing any such change. With the rise of Dom Pedro II[4] (1840-1889), the old aristocracy also remained present and impenetrable, as evidenced by Saraiva (2019):

> During the Second Reign, the traditional system of domination was maintained, this time exercised by Emperor Pedro II, who represented power. Around him, the aristocracy perpetuated its power in a hereditary way, maintaining as much as possible the election and permanence of positions within their family and friendship circles. This governmental aristocracy appropriates the state in order to also command it, further increasing the importance of public office as an instrument of power and a bargaining chip (p. 344).

---

[4] Between the end of Dom Pedro I's reign and the rise of Dom Pedro II, there was the regency period, between 1831 and 1840, when Dom Pedro I returned to Portugal to take the throne and his son, who was only five years old, stayed in Brazil. Dom Pedro II assumed the throne in 1840, after the "coup of majority", at the age of 14 (MOREL, 2003).

Patrimonialist practices went beyond the Empire and continued throughout the governments of the Republic. Even in governments where institutional restructuring was on the agenda, such as that of Getúlio Vargas (1937-1945), in the Estado Novo (New State), with its bureaucratic and dictatorial model. The truth is that there was nothing "new" about it. Despite the Constitutional Charter of 1934, which called for control and standardization when entering the public sector, traditional and dominant power relations remained rooted in the Brazilian state. These relations were transformed into political pacts between the government and the colonels, which is perhaps why there was no change in rural labor legislation, for example (MARTINS, 1994).

In addition, although there was speculation that Getúlio Vargas would break off relations with the landowning colonels, the political leader remained firm in his origins as a cattle rancher and strengthened political pacts. An example of the strengthening of the landowning class was the signing of Brazil's first land grabbing law by Getúlio Vargas in 1931. The decree-law, which was legitimized at the time, dispensed with the need to present purchase/sale documents in order to validate ownership of the property (CASTILHO, 2012).

Also in other governments, such as that of Juscelino Kubitschek (1956-1961), who advocated administrative reform based on the Target Plan. Or that of Jânio Quadros and João Goulart (1961-1964), who promoted significant changes in civil service conditions. In none of these governments was it possible to extinguish the political pact or reduce the presence of clientelist relations and the favoring of agrarian oligarchies.

The same political pact was also maintained during the Military Dictatorship (1964-1985), with the oligarchies ensuring that they participated in decisions during the dictatorial period. Despite the ideological discourse against corruption and the "ghost of communism", which underpinned the military dictatorship and legitimized the persecution and removal from office of progressive parliamentarians, clientelist relationships, possibly corrupt, remained rooted in the structure of the Brazilian state, and even became the military's support base, according to Martins (1994).

> It was from this traditionalist basis that the military drew its political support, ensuring the legitimacy that its regime could obtain from the servile cooperation that was and is typical of the depoliticized and de-ideologized party representation of the oligarchic and clientelist tradition: the mandate is always a mandate in favor of

It is a fact that the problems related to the effective and permanent participation of patrimonialism and agrarian oligarchies in the Brazilian state were not limited to the period of the Military Dictatorship and, in this way, the same principles of Brazilian governance were maintained in the process of redemocratization. An example of this was the government of José Sarney (1985-1990), known as one of the country's most remembered "coronéis" politicians (CASTILHO, 2012). In the Collor government (1990-1992), the focus was on resuming the privatization contracts that had begun during the dictatorial period. In both cases, there was no administrative reform (SARAIVA, 2019).

In the mid-1990s, Fernando Henrique Cardoso took office (1995-2003, over two terms), with the central aim of adopting the minimum state, which had been started by Collor. The administrative reforms of this period followed economic patterns similar to private initiatives, such as the creation of the Ministry of Administration and State Reform and the Master Plan for the Reform of the State Apparatus (PDRAE), centralizing public sector administration (SARAIVA, 2019).

Despite former president Fernando Henrique Cardoso's modernized discourse, he followed the established ways of Brazilian politics: patrimonialism. One of FHC's ministers, Fernando Bezerra, was the nephew of Theodorico Bezerra, a former deputy governor of Rio Grande do Norte and owner of four cotton plantations, one of which covered 14,000 hectares (CASTILHO, 2012).

In the 2000s, President Luís Inácio Lula da Silva took office (2003-2010, two terms). The government in question aimed to create a social inclusion program and guided administrative reforms that considered the quality of public service, as well as expanding transparent processes for hiring civil servants (SARAIVA, 2019).

However, the former president achieved his goals through political alliances with extreme right-wing and/or centrist parties, such as his base made up of politicians from the PMDB (Partido do Movimento Democrático Brasileiro). In addition, the Lula government had the direct presence of representatives of large landowners, as in the case of Geddel Vieira Lima, a cattle farmer and cocoa producer, who was Minister of National Integration in the Lula government (CASTILHO, 2012).

At the end of Lula's term, Dilma Rousseff was elected (2011-2016), maintaining the administrative characteristics adopted by Lula and inserting aspects of modernization into public sector management. In her second term, the President created the Permanent State Reform Commission, integrated into the Ministry of Planning, Budget and Management (MPOG), proposing the reduction of ministries, commissioned positions and secretariats (SARAIVA, 2019).

Former president Dilma governed surrounded by the presence of "colonels" and PMDB politicians, since she formed an alliance with right-wing politician Michel Temer (PMDB) as her vice-president. In her government, patrimonialist relations were also kept alive in the form of agreements with the ruralist caucus. In her second term, former President Dilma had as her strong ally and Minister of Agriculture, Livestock and Supply the politician and large landowner Katia Abreu, known as the "queen of the chainsaw"[5] (CASTILHO, 2012).

The fact is that even in progressive governments like Lula's and Dilma's, there was no legitimate administrative reform that eliminated the presence of clientelist and patrimonialist relations. In fact, before these governments could get elected, they had to build alliances with traditional parties, such as the PMDB (Brazilian Democratic Movement Party) and maintain parliamentary support with the three main parties: evangelical, ruralist and military.

In this way, the patrimonialist relations that have existed since the genesis of the Brazilian state have continued to change and adapt to changes in government. Perhaps they are the only ones that remain even with so many governments.

> Thus, patrimonialism in this period still retained its strength, despite all attempts at reform, undergoing an evolution and adaptation to the new structures formed and established through the institution of the practices of clientelism and favoritism, which still had a lot of weight in the appointment and promotion of public positions, and came to be used as a bargaining chip for personal favoritism. (SARAIVA, 2019, p. 352).

It should also be understood that patrimonialist relations are not natural and anachronistic conditions of the Brazilian state, since they also exist in other countries

---

[5] Available at <https://oeco.org.br/reportagens/28294-rainha-da-motosserra-brasileira-ataca-ambientalistas/> Accessed on: February 1, 2022.

considered central to capitalism, as in the symbolic case of the former presidents of the United States of America, George Bush (1981-1989) and George W. Bush (2001-2009) and are not unrelated to typical capitalist relations. To think that the state structure is totally unnecessary and corrupted by relationships of favors is also dangerous. The fact is that we have a modern capitalist state, merging relations of market interests with patrimonial relations of power. Therefore, it is not enough to understand only the internal relations of the state; we also need to understand the relations of domination of the neoliberal market, which, in a way, conditions the actions of the Brazilian state (SOUZA, 2015).

However, there is no denying that patrimonialism has a strong influence on power relations in Brazilian public administration. The international interference of the market is not in opposition to the corruption of the state, but imposes its actions in conjunction with the internal power arrangements that decide the direction, prospects and capacity for governance of each president.

> In other words, patrimonial domination is not, in the Brazilian tradition, an antagonistic form of political power in relation to rational-legal domination. On the contrary, it is nourished by it and contaminates it. The political oligarchies in Brazil put the institutions of modern political domination at their service, submitting the entire state apparatus to their control. As a result, no political group or party today can govern Brazil except through alliances with these traditional groups (MARTINS, 1994, p. 20).

The patrimonialist relations and the leading role of the landowning class in the Brazilian state are due to the structural condition of this class's alliance with the industrial bourgeoisie, which has maintained the rentier character of Brazilian capitalism as its economic base. This peculiarity of Brazilian capitalism differs from other less authoritarian models of capitalism. Thus:

> As is easy to see, this is not about pre-capitalism or simply late-stage capitalism that depends on what can be called primitive accumulation. For it's not just a question of territorial expropriation, but also of taxation through land rent. The same modern big capital can give up its status as a landowner in the South, but strive to become a landowner in the Amazon. It is, therefore, a different model of capitalism from the classic European or American model: here, the expanded reproduction of capital involves the extraction and realization of land rent. (MARTINS, 1989, p. 80).

In Brazil, as in Latin America, the condition of dependent capitalism has been imposed, which plunders the labor of workers through the overexploitation of labor

(MARINI, 2005), as well as land, using it as a store of value for the generation of capital.

It is these practices of domination and clientelism that determine the patrimonialist characteristics of the Brazilian state, whose control is in the hands of the ruling class in order to enable them to maintain and expand their private assets.

1.2 Capitalist state and the benefits to capitalist agriculture and the maintenance of land concentration

Actions of patrimonialist origin aimed at favoring landowners and encouraging development actions by capitalists have continued in the modern state with new versions, mainly through public policies. And also through the legal apparatus, with environmental flexibilizations, the creation of laws and also through financing projects from national banks, such as the National Bank for Economic and Social Development (BNDES).

The practices of the state "encouraging" capital were not born in Brazil. Since the 1930s, European *Keynesian* principles of an "active" state have been introduced; however, unlike what happened in Europe - an improvement in society's quality of life - in Brazil the state directed investments towards building up infrastructures that would serve the dynamization and territorial fluidity of capital through the National Development Plans (PND) (HESPANHOL, 1999).

Investment policies, territorial infrastructure and the construction of industries were initially concentrated in the Southeast. There was then an economic slowdown in other regions, such as the Midwest. With the aim of spreading regional integration and development policies, the state created Regional Development Superintendencies in the 1970s, such as the Midwest Development Superintendency (SUDECO).

SUDECO was created as a regional planning body, managed by SEPLAN/PR (Planning Secretariat/Presidency of the Republic), whose ideological bias was based on the intellectual formation of technocrats and the participation of dominant agro-industrial capital (ABREU, 2003).

In this way, SUDECO operated as a mechanism to meet the demands imposed by the "pact of interdependence", aligned with the Western bloc, whose objectives were based on maintaining the *status quo* of the ruling class, via ways of making the introduction of international capital more flexible (ABREU, 2003; ABREU, 2008). However, it wasn't just a question of incorporating the regions into the international market; the capitalists also "[...] wanted to turn the lands in the north of Mato Grosso into new assets, in the form of a 'store of value'" (ABREU, 2003, p. 174). Thus:

> The National Project - Brazil Power - envisaged the strengthening of monopoly capital based on multinational industry, which in Mato Grosso would replace subsistence agriculture and polyculture in favor of commercial monoculture to expand the national market, promoting transformations that were consolidated in the imaginary of the 'pact of interdependence'. In this sense, the differences in intervention and results would be determined by the ability to establish a conciliation between the dominant fractions of the class at national and regional level and the interests of multinational monopoly capital, a role played, in the case of Mato Grosso, by SUDECO and, in part, SUDAM - the Superintendence for the Development of the Amazon - as tributaries for promoting the interests of capital in the national territory. (ABREU, 2003, p. 173).

In order to boost SUDECO's efforts, the Midwest Economic and Social Development Plan (PLADESCO) was also created, which instituted the creation of special development hubs, such as the Cerrado Development Program (POLOCENTRO), the Special Program for the Grande Dourados Region (PRODEGRAN) and the Pantanal Development Program (PRODEPAN), the Agricultural and Mineral Hubs of the Amazon (POLOAMAZÔNIA) and the Brasilia Development Program (POLOBRASÍLIA). The programs aimed to organize specific agricultural activities for each area, turning Cerrado areas into monocultures to meet the demand for *commodity* exports, for example.

The northern region of Mato Grosso served as a corridor for land occupation towards the Amazon, as Abreu (2003) points out, "[...] consolidating the Mato Grosso area as a penetration route to the Brazilian Amazon" (p. 176). The advance in land occupation and the intensification of migratory dynamics to the north of Brazil was called the Agricultural Frontier by SUDECO.

In the southern region of Mato Grosso, the poles set up were POLOCENTRO, which sought to encourage areas of pasture, crops and reforestation, with the aim of increasing grain cultivation to 60% and reducing livestock by 40% (TEIXEIRA 2005; HESPANHOL 2000). PRODEGRAN sought to take advantage of the agricultural potential in the Grande Dourados region by growing cereals and oilseeds for export

(ABREU, 2005). PRODEPAN aimed to incorporate the Pantanal area into the national and international market to supply beef (ABREU, 2000).

In 1970, government incentives for the cultivation of grains and oilseeds via PRODEGRAN in Mato Grosso do Sul were part of a large project involving 22 municipalities[6] , with an area of 84,661 km², the center of which was in the "capital of agribusiness", the municipality of Dourados. The boundaries of PRODEGRAN covered six million hectares with satisfactory agricultural capacity (ABREU, 2005).

> This was a program whose resources were made possible through the federal budget, via the Planning Secretariat/Presidency of the Republic. Of the Cr$2,030,000,000.00 (US$250.15 million) programmed for 1976-1979, Cr$430 million (US$52.98 million) would come from non-repayable funds, through the FDAE - Strategic Areas Development Fund -, FDPI - Integrated Programs Development Fund - and the state government and would be earmarked for basic infrastructure and the environment; the remaining Cr$1.The remaining Cr$1,600,000,000.00 (US$197.16 million) would be earmarked for promotional programming and would come from funding bodies such as the then National Economic Development Bank, the National Cooperative Credit System, the Fund for Financing Small and Medium-Sized Enterprises and the National Rural Credit System (ABREU, 2005, p. 161).

The centrality of government incentives for the southern region of Mato Grosso, focused on planting grains such as soybeans and corn (the main exports), gave Mato Grosso do Sul the status of national "granary" after the division process (1979) (MISSIO; RIVAS, 2019). The southern region of Mato Grosso was also forced to deepen the agro-industrialization process (ABREU, 2003).

The creation of agro-industries and the implementation of capitalist agriculture were part of the objectives proposed by the government in question, represented by Ernesto Geisel (1974-1979), a supporter of the tax substitution model who, through Law No. 6,151 of December 4, 1974, released significant resources for the modernization of agriculture and the provision of credits for the implementation of rural companies.

> The aim is to bring the entrepreneurial capacity that has already been shown to develop industry and other urban sectors to the country's agricultural activity. Through the widespread dissemination of rural businesses - small, medium and large - mainly through the

---

[6] Namely: Amambai, Anaurilândia, Antonio João, Bataguassu, Bataiporã, Bela Vista, Caarapó, Dourados, Fátima do Sul, Glória de Dourados, Guia Lopes da Laguna, Iguatemi, Itaporã, Ivinhema, Jardim, Jateí, Maracaju, Naviraí, Nova Andradina, Ponta Porã, Rio Brilhante and Sidrolândia (ABREU, 2005).

government's financial and fiscal support, problems such as inducing the use of projects, taking into account profitability calculations and price stimuli, employing more modern technology and considering the relationship between input and product prices will be better solved (BRASIL. MINISTÉRIO DO PLANEJAMENTO, 1975, p.43).

It was against this backdrop that the National Alcohol Program (Proálcool) was created, with an accelerated increase in the installation of agro-industries in the sugarcane, ethanol and sugar sectors. In the state of Mato Grosso do Sul, sugarcane cultivation currently represents 727,753 hectares, concentrated in the Center-South region, in the municipalities of Nova Alvorada do Sul, Rio Brilhante, Costa Rica, Ivinhema and Angélica, where the agro-industries are also located, according to table 1.

Table 1: Mato Grosso do Sul: municipalities with sugar and ethanol plants and year of start-up - 2020

Location/Municipality	Installed unit	Year of operation	Responsible financial group
Angelica	Usina Angélica	2008	Adecoagro
Aparecida do Taboado	Acoolvale	1983	Unialcool
Batayporã	Laguna Mill	2009	Laguna Mill
Brasilândia	Brasilândia Unit	1979	CBAA (Brazilian Sugar and Alcohol Company)
Caarapó	Caarapó Unit	2009	Raízen
Chapadão do Sul	Usina Iaco Agrícola	2009	Iaco Agrícola
Costa Rica	Costa Rica Unit	2011	Atvos (formerly Odebrecht Agroindustrial)
Dourados	São Fernando Mill	2009	São Fernando Sugar and Alcohol
Fátima do Sul	Fátima do Sul Mill	2011	Fátima do Sul AgroEnergética S.A.
Iguatemi	Dcoil Plant	2002	Centro Oeste Iguatemi Distillery (Dcoil)
Ivinhema	Adecoagro	2012	Adecoagro
Maracaju	Maracaju Unit	1982	Biosev
Maracaju	Vista Alegre Unit	2009	Tonon Bioenergy
Naviraí	Amambai River (former Usinavi)	1983	Amerra Capital Management
Nova Alvorada do Sul	Santa Luzia Unit	2009	Atvos (formerly Odebrecht Agroindustrial
New Andradina	Santa Helena Mill	1978	St. Helena
Ponta Porã	Monteverde Unit	2009	BP Bunge
Rio Brilhante	Eldorado Unit	2009	Atvos (formerly Odebrecht Agroindustrial)
Rio Brilhante	Passa Tempo Unit	1982	Biosev
Rio Brilhante	Rio Brilhante Unit	2008	Biosev
Sidrolândia	Sidrolândia Unit	1977	CBAA (Brazilian Sugar and Alcohol Company)
Sound	Sound Unit	1977	Sound Plant

| Vicentina | Vincentian Unit | 2008 | Central Energética Vicentina |

Source: Baratelli, Nardoque (2021, p. 164).

The period in which the agro-industries in the sugar-alcohol sector were set up, between the end of the 1970s and the beginning of the 1980s, is related to the period in which Law No. 6,151, of December 4, 1974, released significant investments for setting up rural companies. Likewise, the agro-industries set up in the 21st century, especially at the end of the first decade of the 2000s, refer to the massive release of credits for the expansion of capitalist agriculture. The years in which the companies began operating are shown in Table 1.

The fact is that companies linked to the agribusiness sector have strengthened their productive and expansion capacity through credits made available by the public sector, via BNDES (NARDOQUE, 2017). This has led to some Brazilian companies linked to so-called "agribusiness" gaining prominence on the international market, such as Eldorado Brasil, Suzano Celulose e Papel, Sadia, etc.

The exacerbated growth of certain companies, especially in the pulp and paper sector, is also due to government investments in "forest massifs", whose production of wood and charcoal would subsidize the demand of the steel industries and also the advance in pulp production (KUDLAVICZ, 2011).

POLOCENTRO set up specific areas to carry out its activities, dividing them into the states of Minas Gerais, Goiás and Mato Grosso. In Mato Grosso, the following areas were selected: Bodoquena (now part of Mato Grosso do Sul), Xavantina, Parecis and Campo Grande-Três Lagoas (now part of Mato Grosso do Sul). In the Campo Grande-Três Lagoas regions, the area covered by the program corresponded to 1.4 million hectares, along the route of the Noroeste railroad, BR-262 axis (KUDLAVICZ, 2011).

However, the amount of public resources used and the low success of POLOCENTRO shows that the funds were diverted to other purposes than those established by the program. According to Teixeira and Hespanhol (2001), in the municipality of Três Lagoas, for example, the funds were used to clear Cerrado areas, expand large estates and introduce grasses and brachiaria for cattle feed. Just as Abreu shows, with the increase in the percentage of pasture areas.

> In this sense, the program's financing policy distributed the credits increasingly according to the size of the property. The larger the area, the more resources. For example: areas of less than 100

hectares had only 2.2% of the projects approved, receiving only 0.38% of the credit released, while 60% of the projects approved were for properties of more than 500 hectares, which accounted for around 77% of the credit made available. The result of this policy was that from 1975 to 1980, around 915,000 hectares were incorporated into the territory of present-day Mato Grosso do Sul alone, for the exclusive production of soy and beef, with the supremacy of livestock farming. In this unit of the federation, only 6.6% of the incorporated area was occupied by crops in the program areas of POLOCENTRO Campo Grande/Três Lagoas and Bodoquena, while pastures occupied 93% (ABREU, 2003, p.s/n° *apud* KUDLAVICZ, 2011, p. 44).

Despite the fact that the POLOCENTRO's objectives were deviated from, it was during this period that eucalyptus and *pine* monocultures began to be planted, encouraged by the FISET (Sectoral Investment Fund), through Decree-Law No. 1,376 of December 12, 1974. According to Kudlavicz (2011), between 1970 and 1980, 416,653 thousand hectares were planted in the municipalities of Três Lagoas, Ribas do Rio Pardo and Água Clara. With the currency crisis of the 1980s and the drastic reduction in POLOCENTRO's resources, eucalyptus monoculture in the region was reduced.

The recent past of POLOCENTRO's policies in the "eucalyptization" of the Três Lagoas micro-region meant that monoculture plantations returned to the region at the beginning of the 21st century, based on the discourse that there would be a "forestry vocation" (KUDLAVICZ, 2011).

Eucalyptus plantations are concentrated in the eastern region of the state, especially in the municipalities of Três Lagoas, with 263,690 hectares (the largest municipal planted area in Brazil); followed by Ribas do Rio Pardo, with 213,931 hectares; Água Clara, with 131,942 hectares; Brasilândia, with 128,600 hectares; and Selvíria, with 87,321, as shown in table 2.

Table 2: MS: Five municipalities with the largest eucalyptus plantations, in hectares

	2013	2014	2015	2016	2017	2018	2019
MS	651.088	886.381	921.404	993.807	1.117.740	1.117.935	1.124.969
Three Lagoons	140.000	200.000	217.600	230.000	245.000	263.000	263.690
Ribas do Rio Pardo	140.000	175.000	196.000	205.000	210.000	216.000	213.931
Água Clara	95.000	118.000	120.000	125.000	126.000	128.000	131.942
Brasilândia	55.000	58.000	60.000	96.000	120.000	125.000	128.600

| **Selvíria** | 46.000 | 95.000 | 74.350 | 81.500 | 110.000 | 88.000 | 87.321 |
| **Total** | 476.000 | 646.000 | 667.950 | 737.500 | 811.000 | 820.000 | 825.484 |

Source: IBGE Sidra, 2020; Org: The author (2021).

The expansion of eucalyptus monoculture plantations in the micro-region of Três Lagoas was intensified by the demand for raw materials for the factories. Two of the world's leading agro-industries in the cellulose sector have set up in Três Lagoas. This was because, according to Perpetua (2012), Três Lagoas had natural and social conditions.

From a natural point of view, the soil and climate conditions are predominantly tropical; flat terrain, sculpted by sedimentary rocks from the Paraná River Basin, favoring mechanized logging; soils suitable for planted forests and an abundance of water resources, due to the presence of the Paraná River and the Guarani Aquifer (PERPETUA, 2012).

As for the social conditioning factors, at first there was the municipality's history and "forestry vocation"; the low price and availability of land; multimodal transportation infrastructure (road, rail and waterway), as shown in figure 2; a vast source of energy - the "Engenheiro Souza Dias" hydroelectric plant, located in the municipality of Três Lagoas and, above all, what drove the enterprises to the region was the "political will" of the state, which released tax exemptions at federal, state and municipal level (PERPETUA, 2012).

Map 2 - Três Lagoas (MS): location of pulp mills

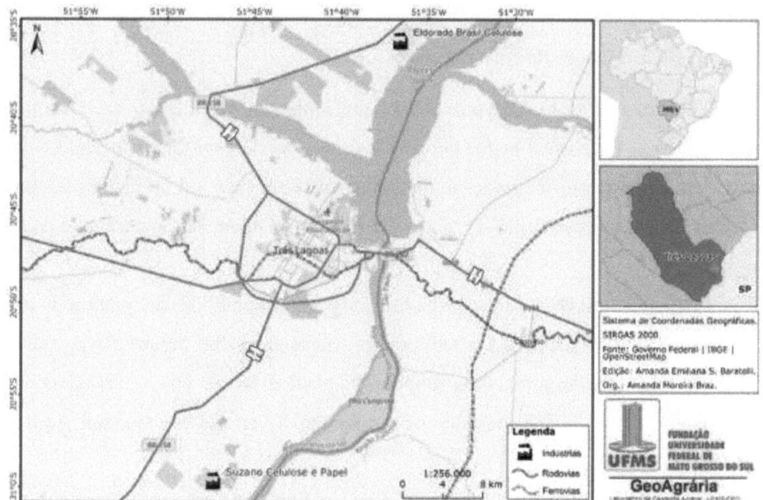

The state of Mato Grosso do Sul used the legal apparatus to attract such companies, such as the enactment of Law No. 093/2001, called "MS Empreendedor", exempting up to 67% of ICMS over 5 years (PERPETUA, 2012). In addition to this percentage of exemptions, a further 23% was added for interstate outbound operations, extending the term of the exemptions to up to 15 years (KUDLAVICZ, 2011).

The legal apparatus also served to facilitate the expansion of eucalyptus, for example through SEMAC/MS Resolution No. 17 of September 20, 2007, which, in its Article 1, exempts environmental licensing for activities involving the planting of exotic species. As the "icing on the cake", the following year, Decree No. 12,528 of March 27, 2008 was enacted, linked to the state's Legal Reserve System (SISREL), which, in its Article 12, allows the restoration of legal reserves through the use of exotic species (KUDLAVICZ, 2011). Thus, deforested Cerrado areas traditionally used for extensive cattle ranching could be reforested with exotic species, such as eucalyptus.

Still from the point of view of exemptions, the municipality of Três Lagoas agreed with the company Fibria (now Suzano S/A) to be exempt from IPTU (Imposto Predial e Territorial Urbano) and ISS (Imposto Sobre Serviços), as well as donating land and infrastructure and waiving around R$50 million in tax revenue, as Perpetua (2012) points out:

> The measures taken by the municipal government have also been
> decisive. As Jurado (2008) shows us, in Três Lagoas the benefits of
> state legislation are added to the exemption from IPTU and ISS, as
> well as the donation of land, earthworks and basic infrastructure in
> the area to set up the industrial unit. Municipal Law No. 1955/2005
> establishes a 5-year tax exemption for investments of up to R$50
> million; 10 years for investments of between R$50 million and
> R$100 million and 15 years for investments of more than R$100
> million. Based on this legislation, according to Kudlavicz (2011),
> the municipal government agreed to waive approximately 50 million
> in tax over 10 years in negotiations with VCP and IP and their
> contractors and subcontractors (p. 168).

The federal government's actions were not only limited to tax exemptions and collections for the public sector, but also financial transfers for the territorialization of the pulp and paper agro-industries, Suzano Celulose e Papel, Eldorado Brasil and International Paper/IP.

The first pulp production company, located in Três Lagoas, began its project in 2006, through an agreement between International Paper/IP and Votorantim Celulose Papel - along with a merger with Aracruz Celulose, giving rise to FIBRIA Celulose S.A., which in 2018 was bought by Suzano Papel e Celulose - and the project cost R$3.88 billion, largely financed by the National Bank for Economic and Social Development (BNDES) (PERPETUA, 2012). The complex was actually inaugurated in 2009, with a capacity to produce 1.3 million tons of pulp (ALMEIDA, 2017).

In 2010, work began on another pulp company in the municipality of Três Lagoas, Eldorado Brasil, a project by the JBS group and MCL Empreendimentos, budgeted at around R$5.1 billion, R$2.7 of which was financed by BNDES and FIP Florestal (PERPETUA, 2012), which opened in 2012. According to Almeida (2017), the company is considered the largest pulp producer in the world, with a capacity of 1.5 million tons per year in a single plant.

The trend towards the *commodification* of agriculture, as in the case of eucalyptus in Três Lagoas, is part of the economic policies of the 1990s, whose precursor was the FHC government (1995-1998 and 1999-2002) during the rise of the neoliberal state model. Foreign indebtedness and the fall in the purchasing value of the Brazilian currency led the country to look for economic alternatives that would maintain fiscal stability and long-term growth, such as the export of raw materials (DELGADO, 2012).

In this sense, according to Nardoque (2017), the FHC government's investments (1995-1998 and 1999-2002) in agribusiness-related activities began with more than R$12.8 billion in his first term and grew exponentially to R$22.4 billion in 2002, the last year of his second term.

State investments in the expansion of export-oriented capitalist agriculture refer to the model of reprimarization of the economy which, according to Veltmeyer and Petras (2014), consists of the exploitation of natural resources and the export of raw materials on a large scale. The model in question intensifies the import-export relationship, via *commodities*, with capitalist centers and enables stability in fiscal balances.

According to Veltmeyer and Petras (2014), the adoption of the economic reprimarization model has been a trend in Latin American countries with progressive governments, because large-scale exports provide a higher percentage of GDP (Gross Domestic Product) and long-term growth potential.

> These are current aspects of capitalism in Brazil and in Mato Grosso that find answers in new/old discourses and in attitudes and actions that are politico-economic, supporting, more than ever, the process of internationalization of the economy, within which sustainable development is consolidated as an ideology, at the same time as public policies are implemented to give vent to the conquests that multinational capital corporations have taken on, even inserting themselves in sectors previously considered strategic and a function of the state, therefore not suitable for private capital. (ABREU, 2003, p. 180).

The Lula government (2003-2010 and 2011-2014), of the Workers' Party (PT), whose bias was based on progressivism, followed the same path adopted by the FHC government, of investing in capitalist agriculture. According to Nardoque (2007), in his first term, the funds allocated amounted to more than R$36.6 billion and jumped to more than R$82 billion in 2010.

The Dilma Rousseff government (2010-2016), also of the Workers' Party (PT), remained aligned with the premise of strengthening capitalist agriculture, since it allocated voluminous amounts of resources to this sector. According to Nardoque *et al.* (2017), the 2015/2016 Safra Plan amounted to R$187.7 billion (one hundred and eighty-seven billion, seven hundred million reais). In addition, his government was branded as "against agrarian reform", because in his term he only made one agrarian reform settlement in Mato Grosso do Sul, in the municipality of Sidrolândia (MS).

It is clear that the rise of former president Michel Temer (2016-2018), through the Political-Juridical-Media-Ruralist Coup that took place in 2016, has maintained the concentration of massive investments for capitalist agriculture, as has happened with the volumes destined for corporate agriculture, totaling R$200.25 billion (Two hundred billion and 25 million reais), according to data from the Ministry of Agriculture, Livestock and Supply (MAPA), in the years 2017/2018. In addition to the voluminous resources for capitalist agriculture, the Temer government also extinguished the Ministry of Agrarian Development (MDA) - the cradle of the main public policies for family farming. It has annihilated public policies for family farming with drastic budget cuts and, along with the right-wing march of the illegitimate government, cases of violence and murder in the countryside have risen, and continue with impunity (CLEPS, 2018; NARDOQUE *et al.* 2018).

The Bolsonaro government (2018-2022) came to power based on consolidated alliances with the capitalist agricultural sector, so it has lived up to its promises and political agreements. Since he took office, around 87.1% of resources for agriculture have gone to capitalist agriculture, while only 12.9% have gone to family farming (ALENTEJANO, 2020). Furthermore, the Bolsonaro government's land agenda, according to the author, has three fundamental characteristics:

> The land policies carried out by the Bolsonaro government in its first year in office have three fundamental hallmarks: (1) the suspension of any land allocation for the creation of rural settlements, indigenous lands and quilombola territories; (2) the creation of mechanisms to free up settlement and indigenous lands for the expansion of agribusiness and other capital interests, such as mining and the construction of hydroelectric dams; (3) the intensification of the legalization of land grabbing, especially in the Amazon. (ALENTEJANO, 2020, p. 365).

In Brazil, therefore, there is the ideological propagation that agribusiness is a highly productive sector, such as the propaganda of the Globo television network, whose *slogan* is "agro é pop, agro é tech, agro é tudo"[7] , constructing the ideology of efficiency, which has served as a justification for the Brazilian state to legitimize huge spending on agribusiness. According to:

> These were the arguments that justified public spending on agribusiness to the tune of R$2.3 billion on the securitization of agricultural debts by the Federal Government between 2007 and 2009, with an estimated annual expenditure of R$800 million on

---

[7] Available at <https://www.youtube.com/watch?v=ZkNhSmE3Cis> Accessed on: July 21, 2021.

Public investment in capitalist agriculture seeks to increase the sector's productive capacity and boost exports in order to increase GDP and balance the balance of payments. In this way, these investments are directed towards export crops, such as soy, corn, cellulose, beef and sugar. These activities have had their cultivation areas expanded through incentives from the special development poles in the Midwest Region since the 1970s and, in recent years, with the neoliberal FHC, Lula and Dilma governments, through the granting of credits.

The growth of sectors linked to capitalist agriculture is an exponential reflection of state investment, as is the case in Mato Grosso do Sul, which in 2002, under the FHC government, received rural credit of R$ 972.6 million. By the end of the Lula government, in 2010, it amounted to more than R$3.4 billion. The main monocultures in the state are soybeans, corn, eucalyptus and sugar cane, concentrated mainly in the Center-South and East regions of Mato Grosso do Sul, as in the case of eucalyptus in the municipality of Três Lagoas - the area under analysis in this study (NARDOQUE, 2017).

In the Brazilian model of capitalism, the relations played out by capitalist agriculture, with the support of the state, boosted the power relations between rentier landowners and industrial capitalists. It was through this fusion between landowners and capitalists that the land-capital alliance pact was made, materialized in the productive latifundia and the territorialization of agro-industries, which made it possible for landowners to earn rent and capitalists to make a profit, based on the reproduction of capital (MARTINS, 1994).

The public fund for maintaining land concentration mediated the relationship between the industrial capitalists and the landowners. In this sense, one cannot consider the merger of the land-capital alliance without first understanding the state's incentives for this "marriage".

The fact is that the Brazilian state allowed the oligarchies to interfere from the outset and consented to the concentration of land when it transferred the administration of wastelands to the power of the states, forming local deals to buy, sell and donate public land. This situation is exemplified by Fabrini (2008) when he discusses the case of southern Mato Grosso do Sul (FABRINI, 2008).

> Thus, the concentration of land in the south of Mato Grosso do Sul is not necessarily the result of the expropriation and amalgamation of small properties in the process of expanding capitalist relations of production. The highly concentrated land structure is due to the process of land occupation, i.e. when the state promoted the transfer/sale of large areas of public land to landowners. Therefore, the land structure in the south of Mato Grosso do Sul [sic] was 'born' concentrated (FABRINI, 2008. p.35).

The appropriation of wastelands - public assets that became the private property of the oligarchies - gave rise to a new business: the sale of land for agricultural colony projects in Mato Grosso. Getúlio Vargas' "March to the West" policy and the arrival of workers from various regions in search of agricultural land made speculation in the sale of land in the region a profitable venture (LENHARO, 1986).

In regions whose economic and population growth was accelerated by the establishment of the Agricultural Colonies, such as Dourados and Ivinhema, due to the large plantations of rice, beans and coffee, land became a commodity for speculation. The dynamics of land speculation grew to such an extent that, in the 1950s, representatives of the state of Mato Grosso denounced the practice. This was because vacant land bought for 5 or 10 cruzeiros[8] became worth 100 to 150 cruzeiros[9] . In Dourados, it was up to 300 cruzeiros[10] (LENHARO, 1986).

> The Secretary [of Agriculture] argues from the perspective of land as a commodity, and its sale as a trade like any other. He goes on to advocate a certain freedom in the granting of plots, a normality that was not surprising. And he cites the case of the granting of more than 100,000 hectares in the Ivinhema region to politicians from São Paulo, when the Land and Colonization Office was set up in Campo Grande. As for the companies' business: 'Condemn modern colonizers because they make a profit on land that they paid for, divided up and demarcated!' (LENHARO, 1986, p. 54).

---

[8] The value of 5 or 10 cruzeiros, in real terms, would be R$2,370.01 and R$4,740.02 respectively.

[9] If the value is between 100 and 150 cruzeiros, the real value would be, respectively: R$ 47,400.02 and R$ 71,100,030.

[10] The value of 300 cruzeiros in real terms would be: R$ 142,200.06. All values have been corrected according to the Central Bank of Brazil's Citizen's Calculator. Available <https://www3.bcb.gov.br/CALCIDADAO/publico/exibirFormCorrecaoValores.do?method=exibirFormCorrecaoValores> Accessed on: January 20, 2022.

Concessions of public land continued, as in the case of the Xingu Indigenous Park project, where several donations were made so that companies could occupy the "empty" territory and prevent the creation of the Park, as Lenharo (1986) points out:

> More than 500,000 hectares were granted to Imobiliária Ipiranga, another to Construções e Comercio Camargo Correia S.A., another to Colonizadora Norte de Mato Grosso, Casa Bancária Financial Imobiliária S.A. (Brunini brothers) and others with intertwined interests, including even capitalists from São Paulo, such as Fúlvio Morganti and others. In 1954, reports the Estado de Mato Grosso, the state government signed 18 contracts for land concessions and companies, each of which never had less than 200,000 hectares. The 10-hectare plots were contiguous and belonged to members of the same group, in order to circumvent constitutional requirements. Colonizadora Norte de Mato Grosso Ltda. alone had reached a probable area of 3,600,000 hectares (p. 54-55).

The problems relating to unequal access to land in Brazil have not been confined to concessions. The state - permeated by patrimonial relations - has found new, more modern alternatives to legitimize land concentration and the formation of latifundia.

From a historical point of view, since the 1970s, with the modernization of agriculture, land has gone through periods of rising and falling prices. However, even with the average price percentage, it has never been accessible to the working class. In situations of crisis and low prices, there has been an increase in land concentration, with the capitalization of income, since "[...] the concentration of land ownership in this process corresponds to an enormous concentration of wealth in general" (GONÇALVES, 1993, p. 12).

The concentration of huge estates should be a problem for landowners, since rural land is subject to Rural Land Tax (ITR). However, in Brazil the contribution to the payment of this tax is derisory, because it refers to the price of the property declared by the owners (GONÇALVES, 1993). In addition, several governments have negotiated million-dollar debt write-offs for the ITR.

In this way, the benefits of land concentration are part of the relations of the patrimonialist state and its actions penalize the majority of the population without access to land and citizenship, as Gonçalves (1993) contributes:

> The national tax system is doubly harmful to the working class, with taxes on income and added value, mainly deducted from wages at source and passed on to the prices of goods and services. On the other hand, property has an insignificant tax burden and, in this

context, land ownership has a negligible tax impact, as does agricultural income (p. 12).

The effective participation of the Brazilian state in maintaining the latifundia can also be seen in the lack of policies to distribute land to workers. Agrarian reform as a policy to change the condition of concentrated land in the country has never been put into practice, even though it is provided for by law.

Agrarian reform projects put into practice in Brazil have always been very timid, even in progressive governments, as in the case of the Lula PT government (2003-2007), whose campaign planned, via the Second National Agrarian Reform Plan (II PNRA), to settle 400,000 new families, in addition to 500,000 land regularizations and the settlement of 150,000 families through the National Land Credit Program (OLIVEIRA, 2011).

However, the target numbers were never achieved. In the case of Mato Grosso do Sul, with inefficient Agrarian Reform policies, only 85 settlements were created for 11,295 families, on 229,085.3 hectares. In the FHC government, 93 settlements were created for 11,782 families, on around 343,996 hectares (NARDOQUE, 2017).

The inefficiency of Agrarian Reform favors policies to maintain the concentration of land, as evidenced by recent data on the land structure of Mato Grosso do Sul, in which rural establishments over 1,000 hectares represent 9.38% of the total and hold 80.45% of the land, while establishments of up to 200 hectares represent 75.22% of the total and hold 0.19% of the land, as shown in table 3.

Table 3: Mato Grosso do Sul: Land structure - 2017

Total area class (ha)	Agricultural Census 2017				
	No. of establishments	%	Area (ha)	%	Average area
0 to minus 200	53.083,00	75,22%	1.509.548	0,19%	28,44
200 to less than 500	6.153,00	8,72%	2.015.063,00	7,27%	327,49
500 to less than 1000	4.718,00	6,69%	3.348.921,00	12,09%	709,82
Above 1000	6.619,00	9,38%	22.286.452,00	80,45%	3.367,04
Total	**70.573,00**	**100%**	**29.159.984**	**100%**	**413,19**

Source: IBGE - Agricultural Census, 2017. **Organized** by the author.

Government investments also served to encourage family farming, but they were very small compared to the amount invested in corporate farming. The FHC government created the National Program to Strengthen Family Farming (PRONAF), with a budget of R$2.3 billion in 2002. In 2010, the Lula government allocated R$11.9 billion to family farming via PRONAF (NARDOQUE, 2007). In addition, other policies were created to provide infrastructure for the Agrarian Reform settlements, such as the Technical, Social and Environmental Advisory Program for Agrarian Reform (ATER) (MATTOS, 2017).

As stated, the governments in question, such as Lula and FHC, adopted clientelist policies of a patrimonialist nature, so as to allocate most of the public resources to capitalist agriculture. In the case of FHC, the government directed investments towards the expansion of economic activities for landowners, with investments of R$22.4 billion for corporate farming in 2002. In Lula's government, R$36.6 billion was earmarked for corporate farming (NARDOQUE, 2017).

The vast financial credits to the capitalist agricultural sector intensified the expansion of monocultures, incorporating land from unproductive estates into the dynamic of leasing for companies to expand their plantations. In the case of Mato Grosso do Sul, this practice created new obstacles to the realization of the Agrarian Reform, since part of the purchase and leasing of properties took place due to the unproductivity of the estates. This situation has paralyzed the expropriation of land by governments.

Furthermore, this incorporation of unproductive land into the dynamics of monoculture plantations has also created an economic block to Agrarian Reform, since the amount of land demanded by the extensive monoculture plantations has led to an increase in the price of a hectare in the region. Baratelli (2019) demonstrates this in the municipality of Três Lagoas, where the price of a hectare rose unparalleled after the expansion of eucalyptus-cellulose. The rise in land prices is an obstacle to Agrarian Reform, since the National Institute for Colonization and Agrarian Reform (INCRA) - with ongoing budget cuts - does not have enough resources to buy farms. Let's take a look at the report published by INCRA in relation to the case of Mato Grosso do Sul:

> Historically, land prices in Mato Grosso do Sul have followed trends in the world economy, especially those involved with traditional agricultural products in the state, such as cattle, soybeans, corn and recently sugar and ethanol.
> The prices of these products have successively surpassed their historical values on the international market and naturally **put pressure on land values, raising them to a level that makes the acquisition of real estate by Incra/MS, via decree 433 (acquisition), almost unsustainable**. The consequence of this is that Incra has little land to offer for purchase and sale, and when it

In practice, Brazil's Agrarian Reform never meant punishing landowners. Unproductive farms were bought by paying the market value, plus any improvements made, via the Agrarian Debt Bond (TDA).

Due to the lack of significant agrarian reform policies to deconcentrate Brazil's land structure, the right of access to land is denied to the peasant and working classes, making it a commodity of privilege for the ruling class. Most of the ownership and control of the land is divided between capitalists and large landowners, who are sometimes confused in the same figure due to the land-capital alliance.

In fact, in the Brazilian case, land ownership is unquestioned by the ability of the ruling class to orchestrate compatible legislation, it is low-maintenance due to low taxation and highly secure, as it is naturally 'protected' without the need for additional spending on private security and has a high return. (GONÇALVES, 1993, p. 14).

In addition to the denial of land to the peasant class, there is also the neglect of policies to recognize indigenous lands (TIs), which are the true owners of Brazilian territory. Under the FHC government, only one demarcation was made in Mato Grosso do Sul - an area where the Guarani Kaiowá have made intense demands for the demarcation of their territory. In the Lula government there were eight demarcations, seven in his first term and one in his last (NARDOQUE, 2017).

In addition to the usurpation of indigenous lands in Mato Grosso do Sul, the race for land in Brazil has been accompanied by the expansion of the agricultural frontier, which is increasingly moving into the Amazon. An example of this was the creation of the Terra Legal (TL) Program, initiated by the Ministry of Agrarian Development (MDA) in June 2009, the intention of which was to regularize vacant land in the Amazon. The initial proposal was to regularize 300,000 titles on land of up to 100 hectares, but pressure from loggers and the enactment of provisional measures allowed the regularization of possessions of up to 1,500 hectares (OLIVEIRA, 2016).

Legal actions that favor land grabbing processes have taken place in various Brazilian governments, such as the Lula government, through Laws 422/2008 and 458/2009, the Temer government with Law 13.465/2017 and the Bolsonaro government with Provisional Measure 910/2019. The debate on the objectives and effects of some of these legal instruments will be explored in greater depth below in topic 2.3.

Finally, we must conclude that the legal apparatus of debt securitization, tax exemptions and financial credits that make it possible to maintain land concentration and the expansion of capitalist agriculture are based on the principles of what is defined in the literature as the patrimonialist state. The patrimonialist forms of politics inherent in the Brazilian state maintain the benefits for those who are part of the public administration, especially those who are part of the dominant classes - in this case, the class of landowners and capitalists.

1.3 Rural taxes as a form of tax amnesty

The use of land as a source of wealth for export to the international market, via *commodities*, shows that financial gains are related to the productive capacity of the territory. However, the concentration of land - unproductive latifundia - through the immobilization of capital in the form of capitalized income escapes this perspective. In this condition, land is a "store of value", capable of earning capitalized income.

The financial gains that occur through the monopolization of captive land are centred on the process of extracting land rent - a social tax (MARTINS, 1981). However, other possibilities of gain are added to this "package", since the state, through tax amnesties, enhances the ability of landowners to pocket money from private land ownership, the result being an increase in their private wealth.

The way in which the state deals with the debts of landowners reveals the patrimonialist relationship present within public administration. According to Oliveira (2001), data released by the Federal Revenue Service in 1994 showed that 59% of landowners with properties between 1,000 and 5,000 hectares evaded paying ITR. Landowners with properties of more than five thousand hectares - huge estates - evade 87% of their taxes.

Even though ITR tax evasion was high during this period, the state enacted yet another bill to facilitate the expansion of capitalist agriculture. These practices can first be seen in the application of Complementary Law No. 87 of September 13, 1996, known as the Kandir Law, the purpose of which is to exempt the export of capitalist agricultural products from taxation, according to Article 3, item II:

> Art. 3 Tax is not levied on: [...]
> II - transactions and services that send goods, including primary products and semi-manufactured industrialized products, or services abroad;

Tax evasion and tax amnesties are part of the privileged reality of Brazil's large landowners, according to the news portal "De olho nos ruralistas" (Keeping an eye on the ruralists), the Oxfam report[11] , and data from the Attorney General's Office reveal that 4,013 individuals and companies owning land have debts of around R$906 billion - more than the GDP of 26 states.

The report also reveals that a select group of 729 landowners have debts of R$200 billion, in properties concentrated by this group corresponding to more than 6.5 million hectares, according to data from the National Rural Cadastre System.

Despite the excessive debts owed by landowners, the Brazilian state is not looking for an alternative to collect them. On the contrary, it has filed provisional measures to amnesty and/or drastically reduce the amounts that should be paid. Former President Michel Temer issued Provisional Measure 733[12] in September 2016 - a month after the coup against President Dilma Rousseff - to reduce rural debts by 60% to 95% for landowners whose rural credit operations were registered at Federal Active Debt (DAU) .

The enactment of Provisional Measure 733, exactly one month after former president Dilma Rousseff suffered the coup, came as a reward to the Ruralist Caucus of the Brazilian Congress for their support in the process, since the coup was orchestrated by various sectors and was considered Political-Legal-Media-Ruralist, in the terms of Nardoque *et al.* (2018).

Although the impeachment of former president Dilma Rousseff was articulated with representatives of the Ruralist Caucus, the PT governments also legitimized

---

[11] Available at: <https://deolhonosruralistas.com.br/2016/12/12/proprietarios-de-terra-devem-quase-r-1-trilhao-uniao/> Accessed on: 23 Jul. 2021 .

[12] Available at <https://blogs.canalrural.com.br/danieldias/2016/09/30/governo-temer-sancionada-mp-733-que-reduz-em-ate-95-algumas-dividas-rurais-veja-se-voce-e-um-dos-beneficiados/> Access: July 23, 2021.

actions that favored landowners. Under the Lula government, Laws 11.763/2008 and 11.952/2009 facilitated the regularization of land grabbed in the Amazon. These laws served as the basis for the creation of the Terra Legal Program which, according to Alentejano (2018), in a survey carried out in 2019, titled 22,523 properties out of a total of 117,179 applications.

The Temer government, as expected, expanded the possibilities for regularizing land grabbing through Law 13.465/2017, which increased the maximum titling area to 2,500ha, in addition to several other facilities included in the law, such as:

> In addition, it established derisory amounts for titling: up to 1 fiscal module, 10% of the value of the bare land; above 1 to 4 fiscal modules, between 10% and 30% of the value; above 4 fiscal modules up to 2,500 hectares, between 30% and 50% of the value. As if that weren't enough, it established easier payment conditions, allowing installments of up to 20 years, with a 3-year grace period and interest of: up to 04 fiscal modules, 1% per year; above 04 to 08 fiscal modules, 2% per year; above 08 to 15 fiscal modules, 4% per year; above 15 fiscal modules up to 2,500 hectares, 6% per year. (ALENTEJANO, 2018, p. 382).

The policies in favor of land grabbing that have taken place under the Temer government are part of the political agreements made to carry out the political-legal-media-ruralist coup, but they also show the former president's personal characteristics. According to Castilho (2012), Temer was involved in a case denounced by the managers of the Cerrado Private National Heritage Reserve (RPPN), in Alto Paraíso de Goiás, in the Chapada dos Veadeiros region. The former president was accused of stealing 2,500 hectares from the 7,000-hectare reserve. The indictment showed that Temer had possession of just 40 hectares, then progressed to 746 hectares and finally 2,500 hectares. The coup plot was to try to double the area by incorporating 2,500 hectares of the reserve in a usucaption action. The attempt was unsuccessful, thanks to the complaint, however, the process was put on hold for years and was only resolved in 2009 and the area of the reserve was recovered.

Jair Bolsonaro (2018-2022) has tried to continue the political project of his predecessors, but this time without any disguise, even using hate speech against political projects for land deconcentration in his campaign, as explicit in his speeches: "As far as I'm concerned, there will be no more demarcation of indigenous land"; "Quilombolas aren't even good for procreating" or with the order made, on the third day in office, for Incra to paralyze[13] all the Agrarian Reform processes underway, he always made clear his appreciation and pact with the Ruralist Bench of the Brazilian Congress.

The Bolsonaro government has continued with its project to dismantle land deconcentration policies and give intense support to Brazilian ruralists. At the very beginning of his administration, according to the Valor Econômico news portal[14] , the president intended to forgive R$17 billion in debts owed by rural producers and agro-industries to the Rural Workers Assistance Fund (Funrural).

There are also episodes in which the Bolsonaro government expresses its favoritism towards the landowning class, but in particular, the president personally supported and pressured the vote on Provisional Measure 910 in Congress - which was not approved due to the vote being postponed beyond the time allowed. The Provisional Measure provided for the regularization of occupied federal land. In other words, it regulates and legitimizes the actions of land grabbers, as Sauer (2019 *et al.*) points out:

> When evaluating MP 910/2019, it is important to note that it continues and deepens what has been projected in recent land regularization measures in Brazil. In December 2016, then-president Temer issued MP 759/2016 to speed up land regularization, which was later converted into Law 13.465/2017. This "land grabbing MP" led to changes in a dozen laws related to land titling and further setbacks during its passage through Congress (Sauer and Leite, 2017; Leite et al., 2018). Although Provisional Measure 910/2019 does not mention Law 13.465/2017 in its amendment, it was enacted in the same spirit, continuing and expanding the possibilities for legalizing the grabbing of public lands throughout the country.

The fact that the Provisional Measure wasn't approved didn't stop the Ruralist Caucus from working together in the Chamber of Deputies. MP 910/2019 was transformed into Bill 2663/2020 which, in its Article 1, provides for occupations on state land:

> Art. 1 This law provides for the regularization of land occupations on land located in areas under the control of the Federal Government or the National Institute for Colonization and Agrarian Reform - INCRA, through the sale and granting of real rights to use real estate. (NR).

> VIII- concession of real right of use: assignment of real right of use, onerous or free of charge, for a certain or indeterminate period of time, for specific land regularization purposes. (Emphasis mine).

The bill also plans to grant titles to families settled in Agrarian Reform settlement projects, but this is not the main objective. According to Sauer *et al.* (2019, p. 2), "The main objective is the titling of large irregular possessions of undesignated federal land, thus another attempt to legalize land grabbing".

While the government fails to get these bills approved, it continues to make life easier for rural landowners. In March 2021, the government reopened the negotiation of Funrural and ITR debts[15] , allowing regularizations and agreements for active debts with the Union.

The Bolsonaro government's amnesties are not limited to taxes and bills, but extend to changes in environmental legislation. The beginning of the process of easing the application of environmental fines was advocated by Michel Temer, through Provisional Measure 867[16] , which exempts landowners from following the requirements of the Forest Code, such as the obligation to reforest deforested areas - which legalized 5 million hectares of illegally deforested land. In addition, the MP provides for discounts of up to 60% on environmental fines - and, according to studies carried out by the TCU (Federal Court of Auditors),[17] only 5% of the fines imposed in Brazil are paid.

The Temer government's policies of amnesty for illegal environmental actions are continuing under the Bolsonaro government. With the expiry of Provisional Measure 867, President Jair Bolsonaro promulgated Decree No. 9.760[18] which, in addition to taking advantage of the flexibilities made by Temer via Provisional Measure 867, also provides for the creation of a "conciliation nucleus", as a form of early amnesty. It also establishes that the offending companies and/or landowners themselves can carry out their own restoration projects.

> "Art. 98: The infraction notice, any terms for the application of administrative measures, the inspection report and the notification

---

[15] Available at <https://www.canalrural.com.br/noticias/governo-reabre-negociacao-para-dividas-do-funrural-e-itr/> Accessed on: July 23, 2021.

[16] Available at <https://g1.globo.com/natureza/noticia/2019/05/29/entenda-o-debate-sobre-a-mp-867-que-altera-o-codigo-florestal.ghtml> Accessed on: July 23, 2021.

[17] Available at < https://www.aosfatos.org/noticias/por-que-o-ibama-arrecada-so-5-das-multas-ambientais-que-aplica/> Accessed on: January 24, 2022.

[18] Available at <http://www.planalto.gov.br/ccivil_03/_ato2019-2022/2019/decreto/D9760.htm> Accessed on: July 23, 2021.

> referred to in art. 97-A shall be forwarded to the Environmental
> Conciliation Center.
> "Art. 142-A. The fine will be converted using one of the following
> methods, to be indicated in each case by the federal environmental
> public administration:
> I - the implementation by the defendant of a service project to
> preserve, improve and recover the quality of the environment, within
> the scope of at least one of the objectives referred to in items I to X
> of the **caput of** art. 140;

The aggravation of the Bolsonaro government's environmental conflict with environmental protection organizations was mainly due to the statement made by the former Minister of the Environment, Ricardo Salles[19] , who, in the middle of the ministerial meeting on April 22, 2020 - already in the context of the pandemic - suggested that the moment of distraction of society should be used to "pass the cattle".

Despite intense protests from society and environmental protection organizations against former minister Ricardo Salles' statement, the minister followed the Bolsonaro government's plan for environmental destruction to the letter. According to the Brasil de Fato news portal[20] , the National Institute for Space Research (INPE) pointed out that fire outbreaks rose by 12% in 2020, with 222,798 recorded. Deforestation hit a record high in the Amazon, increasing by 216% compared to the previous year, according to the Institute of Man and the Environment of the Amazon (Imazon).

Other biomes have also suffered from the ideological alternative of destroying nature in Brazil. Fires in the Pantanal reached 22,119 outbreaks, an increase of 120% compared to 2019. The imposition of the Forest Code on the Atlantic Forest allows predatory actions to penetrate the few remaining remnants of forests.

The atrocities involved in the destruction of the Amazon were mediated by illegal alliances with groups of miners and loggers, who plunder the traditional territories of the peoples of the Amazon. Notices of violation of the Forest Code fell by 34%[21] , because the bodies responsible for inspection were dismantled, as was the

---

[19] Available at <https://g1.globo.com/politica/noticia/2020/05/22/ministro-do-meio-ambiente-defende-passar-a-boiada-e-mudar-regramento-e-simplificar-normas.ghtml> Accessed on: July 23, 2021.
[20] Available at <https://www.brasildefato.com.br/2021/04/21/ricardo-salles-13-fatos-que-fazem-do-ministro-ameaca-ao-meio-ambiente-do-planeta> Accessed on: July 23, 2021.
[21] Available at <https://oeco.org.br/noticias/sob-bolsonaro-autuacoes-do-ibama-sao-as-menores-em-uma-decada/> Accessed on: January 20, 2022.

case with the Chico Mendes Institute for Biodiversity Conservation (ICMBio) and the Brazilian Institute for the Environment and Renewable Natural Resources (Ibama).

The illegal practices carried out by former minister Ricardo Salles culminated in an investigation by the Federal Police, in which the former minister is accused of taking part in a scheme to illegally export timber to the United States and Europe, according to the G1 news portal[22] . Due to pressure over this, former minister Ricardo Salles was dismissed by President Bolsonaro and replaced by Joaquim Álvaro Pereira Leite, who is now tasked with continuing the government's project of destruction.

In this way, the use of rural tax amnesties, draft laws and actions that make environmental legislation more flexible serve as tools for maintaining the order of the patrimonialist state, which continues to benefit the representative fractions of land owners who are part of the public administration. Furthermore, it reinforces the understanding that land, whether productive or unproductive, serves as a source of monetary income, since in rentier capitalism land is a source of income and profit, and therefore of capital production.

---

[22] Available at <https://g1.globo.com/df/distrito-federal/noticia/2021/06/23/ricardo-salles-entenda-operacao-contra-exportacao-ilegal-de-madeira-que-mira-ministro-do-meio-ambiente.ghtml> Accessed on: July 23, 2021.

CHAPTER 2: A NEW STATE IN AN OLD FOUNDATION STRUCTURE: THE PROCESS OF THE DIVISION OF THE STATE OF MATO GROSSO

The long narrative of the divisionist movement for the fragmentation of the state of Mato Grosso is a source of pride for part of the population of southern Mato Grosso. However, when it comes to analyzing the territorial formation in some depth, one discovers how tragic it was, since it was based on the individual interests of the elite that controlled the south and north of Mato Grosso.

The central motivation for the divisive movement in the south of Mato Grosso was the desire to control a single state that would generate class identity and overcome the internal divisions of the local bourgeoisie. The strong presence of cattle ranching in what was then southern Mato Grosso and its rapid growth, with the constitution of an individual elite, with little and/or no relationship with the capital, Cuiabá, was the powder to ignite the divisionist flame.

The division in 1977 gave the population the false impression that there would be a new state, whose political and economic project would also seek change. However, in practice, the division of the state only gave the elite in the south of Mato Grosso the chance to govern and guide their interests, because the land-owning structure and the presence of traditional groups in power have not changed, quite the contrary, they are still present today.

According to the literature on the subject, the territorial formation of what is now Mato Grosso do Sul was based on donations/concessions/purchases of vacant land. There was also violence, banditry and colonelism. This is how the traditional oligarchic - and political - groups set up their estates. In the territory of Mato Grosso do Sul, the population did not (and still does not) enjoy the right to land for the reproduction of life.

This is how the state maintains its highly concentrated land structure, with the presence of a few municipalities, many large estates and various obstacles to peasant recreation. The concentration of land present in Mato Grosso do Sul extends to all regions of the state, as in the case of the Bolsão, in the eastern region, a territory where latifundia reign and do justice to Mato Grosso do Sul's land structure.

2.1 Division of the state of Mato Grosso and the oligarchic legacies in Mato Grosso do Sul

The strong presence of land concentration in Brazilian territory is part of the historical relationship between the use of the state to fulfill private interests. To this end, illegitimate strategies such as land grabbing - based on state and notary corruption - or donations and grants of possession of vacant land, based on clientelist relations, are important chapters in the essence of how huge latifundia were created.

Mato Grosso do Sul's land structure was no different. In this sense, in order to understand the territorial formation of the state of Mato Grosso do Sul, based on its history of occupation and the land concessions made by the government, it is necessary to return to the processes that preceded the division between Mato Grosso and Mato Grosso do Sul.

At the outset, it is worth pointing out that the former state of Mato Grosso had a vast territorial dimension and that economic relations in the north and south were always different. Due to the difference in regional economic activities, the interests of the ruling class also differed. The similarity between fractions of the same bourgeoisie and landowners, with only regional distinctions, was centered on the aim of expanding latifundia and concentrating land.

Motivated by the expansion of the frontier to the west, the occupation of the province of Mato Grosso had its origins in the bandeirantes' journeys to the interior of the country in the mid-17th century (TEIXEIRA, 2009). The routes taken by the bandeirantes were intended to explore new areas, especially those with potential for exploitation. The bloody journey of this march was responsible for enslaving and decimating indigenous peoples (SILVA, 2014).

The advance of the bandeirantes' entourage into the province of Mato Grosso was responsible for the discovery of gold in the 18th century, in the region where the city of Cuiabá was later located. The exploitation of the gold deposits in Mato Grosso became the first and main economic activity, an attractive source for the origin of the occupation processes.

Between the middle of the 18th century, the state of Mato Grosso went through a systematic economic crisis, which converged with problems of various kinds. The economic problems were due to the decline of mining and the removal of the

Portuguese who administered the captaincies, as a result of Brazil's independence (1822). In addition to these crisis points, the province also faced political problems, such as the occurrence of the "nativist movement", through the Rusga, or Cuiabana Rebellion (1834), according to Alves (2017):

> Since the end of the 18th century, with the crisis of the gold economy, the region's development was marked by material decay, and was especially marked by the displacement of capitalists and slave labor beyond the boundaries of Mato Grosso. It is crystal clear that this displacement was the result of the need to demobilize capital from mining activities to be used in other branches of production. Brazil's independence also meant that Portuguese officials who were responsible for the administration of the former captaincy had to leave. Finally, the situation of decadence was aggravated by the Rusga, or Cuiabana Rebellion (1834), a "nativist movement" that turned against the then dominant Portuguese commercial bourgeoisie: merchants were killed and persecuted, commercial houses were looted, and those who managed to stay safe fled the province. (ALVES, 2017, p. 12)

In addition to these problems that were already affecting Mato Grosso, the outbreak of the Paraguayan War intensified the problems, especially those of a political nature. According to Corrêa (2006), the Paraguayan War led to the appearance of the first phenomena of banditry in Mato Grosso. The practices of banditry were essentially based on the violent practices of gunmen, which generated total lack of control to contain the cases, mainly because of the vast territorial dimension of the state.

The violence carried out by bandits in Mato Grosso intensified the dispute for power of command between the state's colonels. The colonels incorporated banditry as a form of confrontation in party-political struggles. For Corrêa (2006), the incorporation of these practices into the local political dispute led to the region being known as a "lawless land", whose only law to be followed was the 44 caliber:

> It was from this moment that, in parallel with the coronelista phenomenon and also as a consequence of the domination of the colonels, a banditry unprecedented in Brazilian history developed in Mato Grosso. The region of Mato Grosso then came to be known as a land without law, or where the only existing law was article 44, or the law of the 44 caliber (CORRÊA, 2006, p. 37).

The problems that occurred in Mato Grosso happened concomitantly with the occupation of the province, which also expanded to the south, due to the flexibilization of economic activities. Gold mining was replaced by the introduction of rubber and poaia and, above all, by the cultivation of sugar cane and livestock (MORENO, 2007).

In the southern region of Mato Grosso (present-day Mato Grosso do Sul), non-indigenous occupation took place in combination with extensive cattle ranching and the expansion of yerba mate cultivation by the Matte Laranjeira company after the end of the Paraguayan War (1864-1870). The company began its activities at the end of the 19th century, still in a modest way and, as it expanded its cultivation areas, it became a powerful economic group. The company's growth was related to the amenities granted by the state, since, via Decree Law No. 8799 of December 9, 1882, the company secured the concession to exploit the mate herbs on vacant state land (MORENO, 1994).

Due to the rapid growth of Matte Laranjeira, Fabrini (2008) considers that, together with the unproductive latifundia, masked by huge areas of livestock, the cultivation of yerba mate contributed in part to land concentration in the southern region of Mato Grosso. This is because the monopoly actions in the cultivation of yerba mate led to the development of small properties, mainly because the company concentrated large areas and the infrastructure for transporting production, which forced farmers from small properties to sell their produce to the company.

The 19th century occupation movement in the southern region of Mato Grosso was based on territorial disputes over land ownership, whether by families from São Paulo or Minas Gerais who, shielded by government documents, invaded areas that already belonged to their rightful owners, or by indigenous communities (BITTAR, 1999). And as Borges (2012) shows, the same happened in the case of Sant'anna do Paranahyba, with the enslavement of the Cayapó, according to:

> Through the savannahs and mountain ranges, varadouros, banks, roads and monsoon rivers that connected the Captaincies of São Paulo and Mato Grosso, it is possible to glimpse encounters and mismatches, materialized in contact with those who occupied large and small tracts of land and the indigenous nations. These peoples saw their lands and rivers taken over by the groups or individuals who arrived in those parts, mainly from São Paulo and Minas Gerais (BORGES, 2012, p. 109).

The occupation of the south of Mato Grosso also took place in various regions, such as where the Pantanal is today. More to the southwest of the state of Mato Grosso, the invasion of the region took place at the same time as the migratory flow of people from Minas Gerais and São Paulo to raise cattle, as happened in the Sertão dos Garcia Leal, who entered through Sant'anna do Paranahyba - today's eastern Mato Grosso do Sul (LEONARDO, 2020).

Conflicts over land ownership continued to occur, but this time between oligarchies who wanted to expand their estates. The comitivas from Rio Grande do Sul were looking for vast tracts of land to introduce extensive livestock farming (TEIXEIRA; HESPANHOL, 2006). In order to acquire the land, the Gaucho entourage had to confront "[...] the unbridled monopoly of Matte Laranjeira, which threatened to become a 'State within the State' by occupying a large part of the territory in the former south of Mato Grosso" (BITTAR, 1999, p. 95).

According to Bittar (1999), the embryo of divisive interests came about because of the state's extensive monopoly and concessions to the Matte Laranjeira company, since the Gaucho entourage and other landowners in the state felt aggrieved. Thus:

> But what reasons triggered separatism at the end of the 19th century? Basically the monopoly of the Mate Laranjeira company, which prevented the settlement of southern Mato Grosso. Manoel Murtinho, former president of the state and a partner in the company, even stated in a threatening letter in 1907 that he would rather see the region occupied by foreign companies than by immigration from Rio Grande do Sul. According to Arlindo de Andrade, Mate Laranjeira, a company that *'ordered and didn't ask'*, became a kind of *'state within the state'*, slowing down the settlement of southern Mato Grosso. Those who rose up against his power and unbridled latifundia were immediately branded as *'agitators'*, *'sons of other states'*, who formed an *'organized committee'*, guided by an *'audacious plan to assault this state patrimony, entrusted to our custody'*. Such was the content of a document issued by Mate in 1931! (BITTAR, 1999, p. 97, emphasis added).

The predatory advance of the Matte Laranjeira Company's power caused discomfort among other landowners who began to pressure the state of Mato Grosso to cease granting land to the company. In 1916, the state ended the company's land concession contract, but the company remained influential and participatory in government actions until 1943 (LEONARDO, 2020).

The strength of the Matte Laranjeira company in occupying huge estates on state land continued even after the Proclamation of the Republic (1889), which transferred control of the land from the federal government to the states (FABRINI, 2008). The dispute between the company and local landowners continued for 27 (twenty-seven) years after the change in Brazilian land legislation, with the concession contract only ending in 1916.

Like this:

> With the proclamation of the republic, land policy became the
> responsibility of the states. The republic transferred power over land
> to the regional oligarchies, who began to decide on its ownership
> within the state domain, monopolizing its possession and putting
> into practice the policy of concentration. In this context, vacant state
> land was transferred through sale and lease to large farmers and
> capitalist companies operating in the sector (FABRINI, 2008, p. 60-
> 61).

The power of decision over land, transferred to the regional oligarchies, culminated in sales/donations of vacant land to landowners. The donations, as a form of gift, were based on clientelist relations between the landowners and the oligarchies that were part of the public administration. According to Moreno (1994, p. 100):

> [...] it represented a prize for the regional oligarchies, more
> specifically landowners and large squatters, who wanted to see the
> power to decide on the fate of vacant land decentralized. The ruling
> classes in the states would then be able to influence the distribution
> of land more directly, according to their economic and political
> interests.

Since the transfer of land control to the states, the Mato Grosso government, through the state's first land law (Law No. 20/1892), has favored access to portions of the territory for landowners, individual capitalists, agricultural companies and colonization companies (LUIZ, 2020). Thus:

> The state's first land law (Law No. 20/1892) and its regulation
> (Decree No. 38/1893) guaranteed the regularization of consolidated
> occupations, sesmarias and possessions until 15/11/1889, thus
> changing the 1854 deadline established by the Land Law of 1850.
> They also ensured the right of first refusal to buy vacant land that
> was under private ownership, whose titles did not meet the
> requirements for legitimization or revalidation (MORENO, 1999, p.
> 68).

Although there were loopholes in the law, such as the preference for buying vacant land from those who owned it, the local oligarchies who controlled the state of Mato Grosso used the legal instrument to further their interests. In this way, they allowed themselves to legitimize large landholdings in irregular situations, since the sales would serve to fill the state's coffers. Therefore:

> The data shows that the exorbitantly large tracts of land were still
> the result of occupations or possessions, which would have fallen
> into escrow if the deadlines hadn't been constantly extended. The
> various rulers preferred not to get politically involved with the
> landowners and, on the other hand, kept the revenue collected for
> the state treasury high (MORENO, 1994, p. 111).

The land law (No. 20 of 1892) of the state of Mato Grosso served as an instrument, from a legal point of view, to legitimize illegal possession and land grabbers in the state. By 1897, 3,203 possessions had been registered in the state (LUIZ, 2020). According to Moreno (2007):

> [...] of this total, 1333 were possessions prior to 1854, the deadline imposed by federal regulations for the registration of possessions and sesmarias; 1393 possessions were prior to 1899, the deadline imposed by state regulations for registration; and 297 possessions had no date of occupation. Of the total number of possessions, only 452 had been measured and demarcated and only 1,941 had declared the area occupied, with an area of 13,753,011 hectares (MORENO, 2007, p. 68).

In addition to legitimizing irregular possessions, provisional titles were also granted between 1899 and 1902 in the state of Mato Grosso by the government of Antonio Pedro Alves:

> [...] 104 provisional titles, covering 23,639.512 hectares. In the same period, it issued 58 definitive titles covering a total of 125,749.50 hectares, of which 85,834.50 were excess. In other words, only 31.75% of the legal area. With regard to private land, i.e. possessions, 201 titles were legitimized, covering 3,051,280.75 hectares, of which 2,942,559.5 were legal and the rest excess. The average area corresponded to 15,180.501 hectares of land for each legalized title. [...] These records referred to the finalization of the measurement and demarcation processes of the vacant lands that had already been alienated, as a result of sales or legitimization of possessions. (MORENO, 2007, p. 69).

As the end of the yerba mate empire was announced, land concessions were transferred to landowners, cattle ranching expanded on the estates and became the main activity in southern Mato Grosso. The proximity of the south and east of Mato Grosso to the Southeast region and the possibility of exporting "standing cattle" via the Port of Corumbá and the Noroeste Brasil Railway, linked from Campo Grande to the state of São Paulo, meant that the economic dynamism, combined with other regions and centered in the south of Mato Grosso, overruled the capital Cuiabá - which at the time was responsible for political control relations (HESPANHOL, 2000).

The dissatisfaction of cattle ranchers in the south of Mato Grosso with the political control of Cuiabá intensified the divisionist agenda. Among the demands, the main one was the search for political control of the "new" state, since the economic relations of the southern Mato Grosso oligarchies were developing in Campo Grande, with the growth of an intellectual elite, the result of the sons and grandsons of the ranchers who had become "doctors" with studies in other states (BITTAR, 1999).

Despite the intellectual bias of the advance of the divisionist agenda, incorporated by the elite of southern Mato Grosso, through the sons of farmers who went to study in São Paulo and Rio de Janeiro, the bias of the struggle for the division of the state had its essence in coronelista disputes, based on the objective of containing the state's power of command:

> [...] decisively influence the direction of republican politics, which was then run according to the interests of the coronelista groups that alternated in the state's political command. In this sense, the state's political context was characterized by constant struggles between the colonels, which transformed the state of Mato Grosso into a republic of colonels (CORRÊA, 2006, p. 59).

The divisionist movement began to put pressure on then-president Getúlio Vargas, but the government's "nationalist" bias, the agenda of national integration and the settlement of border regions, meant that national efforts were directed towards other projects. Such as the "March to the West", the precursor to the creation of the National Agricultural Colonies (LEONARDO *et al.* 2021).

The implementation of National Agricultural Colonies took place in 1938, via the Brasil Central function, with the consolidation of colonies in Ceres (GO) and Dourados (MT) - currently in the south of MS -. The installation of the Colonies, in fact, served to populate the region, in addition to intensifying other forms of agriculture, with the planting of foods such as rice, cotton, peanuts, coffee, beans, manioc and castor beans (LEONARDO, 2020).

Once the settlers' areas were definitively titled, some plots were sold. The sale led, in part, to land reconcentration and speculation in property prices, since, due to the mass installation of settlers, the state had invested in infrastructure to integrate the regions (HESPANHOL, 2000).

The creation of agricultural colonies also became an initiative of private companies, which used the discourse of the need to bring in workers in order to obtain free land from the state. In this way, huge areas were given over for the creation of colony projects that failed, but the land was not returned to the state (ABREU, 2001).

> This is one of the many episodes of fraud in Mato Grosso, because it was a contractual clause that if the company that acquired the land didn't implement the projects within five years (until 1978,

> therefore), the land would revert to the government. Among the
> companies that acquired land were: Rendanyl (later Otsar) 1 million
> hectares; Indeco (Ariosto da Riva) 400,000 hectares; Colniza
> (Lunardelli group) 400,000 hectares; and Juruena (João Carlos
> Meirelles) 200,000 hectares. Of these, only Indeco implemented a
> colonization project in good time. The others, on the other hand, did
> little or nothing. However, the lands were not returned to public
> ownership (OLIVEIRA, 1996, p. 146-147).

In this way, it can be seen that, from the beginning, in the occupation and consolidation of the large estates, the state was the fundamental mediator, since it always participated, via material and political subsidies, in the maintenance and growth of the regional oligarchies. The constitution and strength of the landowners in the southern region increasingly intensified the divisionist agenda, which continued to hover over all the governments.

The advance of agricultural colonies, due to the vast occupation of the southern region of Mato Grosso; immigration from Rio Grande do Sul to raise cattle; the growth of properties, due to the increasing export of cattle to various regions of Brazil, kept the economic revenue of the state of Mato Grosso centered in that region. Thus:

> In fact, from the 1930s to the 1960s, the supremacy of the south was
> reaffirmed in various works expressing the southern cause: Arlindo
> de Andrade, for example, in **Erros da Federação**, from 1934,
> already pointed out, after analyzing the economic situation of Mato
> Grosso, that: '*The state lives today on what the south yields*'. Emílio
> Garcia Barbosa also stated that '*the revenue from the south amounted
> to more than two thirds of the total*'. Oclécio Barbosa Martins, in a
> study on the state's geopolitics, concluded: '*The south has everything
> but administration*'. (BITTAR, 1999, p. 101, emphasis added).

The economic revenues of the south and the control of the elite of southern Mato Grosso were centered in Campo Grande, so there was no relationship with Cuiabá other than institutional political control. Because of these facts, with each growth of the southern region, the divisionist movement heated up.

In this way, the elite in the south of Mato Grosso began to include issues relating to culture and nature in their divisive agenda. From the point of view of culture, according to them, the southern region had much more cultural ties with the Southeast and South than with the north of Mato Grosso. With regard to natural issues, the northern region was much more similar to the climate and the Amazon biome, while the south was similar to the southeast (SILVA, 2006).

In fact, all the insertions of reasons justifying the division of the state were part of an individual political project by the elites of the south of Mato Grosso, who aimed to constitute individual - or group - power in the state. The population in general was never involved in the division movement, except for the 20,000 signatures taken to the 1934 Constituent Assembly, attached to the divisionist agenda (SILVA, 2006).

Despite the many justifications and demands, the division of the state was only consolidated in 1977, during the military dictatorship. This was because, according to Bittar (1999), the idea of national border protection from the "communist ghost" that haunted President Ernesto Geisel served as the basis for establishing a government of its own for the south of Mato Grosso, thus making it possible to be more careful about guerrilla movements in neighboring countries.

In 1977, through Complementary Law No. 31, the state was divided into Mato Grosso and Mato Grosso do Sul. Despite the territorial division, nothing changed politically. The agrarian oligarchies remained in power more than ever, with the strong presence of the traditional parties inherited from Mato Grosso, such as the UDN (National Democratic Union) and the PSD (Social Democratic Party).

In addition to maintaining patrimonial state relations, based on clientelism, they also established the continuity of the power of traditional Mato Grosso families in Mato Grosso do Sul. For example, the Corrêa da Costa, Müller, Ponce and Barbosa Martins families are among the most influential in Mato Grosso do Sul. Therefore:

> Some of these elites have been dominant in Mato Grosso politics for a long time, sometimes as families. The oligarchies took turns holding elected or appointed positions, some of them notorious, such as the Corrêa da Costa family, who had held public office since the Empire. Another part of this family was the Müller family, made up of Filinto Müller, chief of police of President Vargas' federal district, as well as his brothers Fenelon, Júlio Müller and the latter's brother-in-law, João Ponce de Arruda. (ARRUDA, 2019, p. 77).

The Corrêa da Costa family is made up of descendants of the Portuguese Francisco, whose sixth son Antônio was president of the province of Mato Grosso and passed on the lineage to his sons, especially Antônio Corrêa da Costa and Pedro Celestino, who, in the Republic, were presidents of the state. The generation continued to be involved in politics, this time with Pedro Celestino's son, Fernando Corrêa da Costa, who became influential in Mato Grosso and Mato Grosso do Sul (ARRUDA, 2019).

The Corrêa da Costa family, important in the political history of Mato Grosso and Mato Grosso do Sul, were also, from an economic point of view, owners of large tracts of land, which guaranteed them power and influence in political decisions (ARRUDA, 2019).

The political relationships that mark the history of Mato Grosso and Mato Grosso do Sul are based on a few families who were related to each other and favored each other politically, as in the case of the link between the Corrêa da Costa family and the Müller family.

> Fenelon Muller graduated as a civil engineer in 1918 and was appointed to work for the Noroeste railroad, another institution that provided various resources. According to Ribeiro (s/d), the appointment was due to the influence of Pedro Celestino, his relative. In Três Lagoas, he took part in the construction of the bridge over the Paraná River and was appointed mayor of the city by Governor Pedro Celestino in 1924. Governor Mário Corrêa da Costa, his second cousin, appointed him mayor of Cuiabá in 1927, a position he held until 1930 (ARRUDA, 2019, p. 173).

Still on the subject of the oligarchic and allied families of Mato Grosso and Mato Grosso do Sul, the Muller family was also related to the Ponce family, known for being part of the Brazilian political elite, through Generoso Ponce. The Ponce family, like Muller, favored clientelist relations in politics, as Arruda (2019, p. 178) points out: "Ponce Filho's insertion into the ruling class was partly due to his being the son of the legendary politician, but mainly due to his family relations with the Mullers."

The continuity of conservative political relations, even after the division of the state, reveals that the territorial fragmentation project was based on the particular interests of the dominant elites in controlling power. According to Arruda (2019), 47.5% of the politicians who made their careers in the southern region of Mato Grosso were born and belonged to the Cuiabá elites.

In addition to family relations, the high level of presence of political elites from the north of Mato Grosso in the southern region was due to alliances of interests. In the case of former governor Wilson Barbosa Martins, a descendant of one of the first settlers in the southern region - José Francisco Lopes - the political pacts were aligned with the positions of the Müller families and, later, in coalition with Fernando da Costa Corrêa, to support the candidacy of Pedro Pedrossian, representative of the UDN (BITTAR, 1999).

Despite the movement to divide the state, the politicians who stood out among the elites of Mato Grosso continued to maintain their political and family power in Mato Grosso do Sul. The politicians who stood out, with several mandates in state government, city hall and/or representation as deputies and senators, were loyal to the interests of the UDN and PSD, responsible for representing the ruralist caucus, as in the case of Wilson Barbosa Martins, Arnaldo Estevão de Figueiredo, Fernando Corrêa da Costa, João Ponce, Fenelon and Filinto Müller.

In this sense, the effort to divide the state between fractions of Mato Grosso's political class served to facilitate the materialization of specific interests, especially with regard to the advance and consolidation of large farms. The economic growth of the southern region, the rise in the price of land - due to the establishment of Agricultural Colonies and the advance in the sale of cattle in eastern Mato Grosso - attracted the interests of politicians who were also landowners. Thus:

> Governor Estevão A. Corrêa was the grandson of Cesário C. da Costa, son of Antônio and Maria da Conceição. According to Frank (1999), Estevão had 40,000 hectares in Miranda, the entire municipality comprised 60,454 hectares, he had 84,215 hectares in Aquidauana and 40,161 hectares in Três Lagoas (ARRUDA, 2019, p. 188).

The division of the state maintained the essence of the interests of the Mato Grosso elite, who began to unfold their perspectives in Mato Grosso do Sul, where the dominance of family political pacts would be homogeneous. Thus, with the end (from the point of view of elections) of the aristocracy between Corrêa da Costa, Muller, Ponce and Barbosa, the political groups began to organize themselves to nominate their successors in order to maintain their personal interests in force, as happened with the election of Pedro Pedrossian (1980-1983 and 1991-1995), from the UDN, as evidenced by Bittar (1999):

> [...] state power has so far been under the absolute control of the 'heirs' of Filinto Muller (PSD) and Fernando Corrêa da Costa (UDN). What's even more interesting is that the UDN and PSD left stronger roots in the part of Mato Grosso that broke away, demonstrating that the forces of continuity were more consolidated in the 'south' than in the 'north', in other words, in that part that was calling for renewal (BITTAR, 1999, p. 113).

Pedrossian, appointed by former governors and representatives of the PSD and UDN, was given the task of representing all the interests of the landowning class - of which he was also one. In this sense, the historical alliances of the political pact between the traditional representatives of the state were always present in the constitution of the "new" state political power.

It is in this sense that the da Costa Corrêa family is keeping its reach into southern Mato Grosso politics alive, this time represented by the granddaughter of Fernando da Costa Corrêa - Tereza Cristina da Costa Corrêa Dias, who joined the Brazilian Social Party (PSB) and was elected Federal Deputy in 2014 and leader of the PSB caucus in the Chamber of Deputies. In 2017, Tereza joined the Democratic Party (DEM) and became leader of the Ruralist Caucus and was one of the main people responsible for the approval of Bill 6.299/2002[23] which facilitates the regulation of pesticide registrations in Brazil. Tereza Cristina (DEM) is currently Brazil's Minister of Agriculture, Livestock and Supply in the Bolsonaro government.

The former Minister of Health in the Bolsonaro government, Luiz Henrique Mandetta (DEM), also has his origins in the political and oligarchic elites of Mato Grosso do Sul. Mandetta is the son of Hélio Mandetta, the former mayor of Campo Grande. He also has relatives with notable involvement in the state's politics, such as Senator Nelsinho Trad, Federal Deputy Fábio Trad and the current Mayor of Campo Grande Marquinhos Trad. Mandetta began his career in politics as a Federal Deputy, for the Democrats (DEM) party, and maintained his political alignment with the interests of the extreme right, maintaining a position of opposition to the Dilma Rousseff (PT) government and voting in favor of her *impeachment*. He also voted in favor of the public spending ceiling PEC and the Labor Reform, among others.

The maintenance of elites and the formation of new groups are also based on the performances of traditional predecessors, as in the case of the Tebet family. Ramez Tebet (former mayor of Três Lagoas and former governor of Mato Grosso do Sul) shows this heritage and admiration when he says in the opening of the second edition of his book "Santana de Paranaíba" (2002): "I grew up among the descendants of the Garcia Leals, listening attentively to their stories (an example of courage, honesty and work)" (p. 9). The Tebet family has become one of the most traditional families

---

[23] Bill no. 6.299/2002. Available at
<https://www.camara.leg.br/proposicoesWeb/fichadetramitacao?idProposicao=46249> Accessed on: January 5, 2022.

involved in politics in Mato Grosso do Sul and Três Lagoas (now with their daughter Senator Simone Tebet).

Both Senator Simone Tebet (MDB) and Minister Tereza Cristina (DEM) had their campaigns financed by a large amount of money from companies located in the Bolsão (MS). Fibria S.A. (now Suzano), a pulp and paper company based in Três Lagoas, and Iaco Agrícola, a sugar and alcohol company based in Chapadão do Sul, spent R$5,262,844.00 (five million, two hundred and sixty-two thousand, eight hundred and forty-four reais) on the campaigns of 18 candidates, including Simone Tebet and Tereza Cristina (MELO, 2021).

Family inheritance as colonizers, pioneering territorial formation and control as large landowners and interference in state policy materialize the increasingly strong continuity of the genesis of the Brazilian state, whose character is patrimonialist. And the ideology that there is a fine legacy of politicians who were forerunners of Brazilian territorial formation has been built up in part of the imagination of Brazilian society. An example of this, in the case of Mato Grosso do Sul, can be seen in the tributes to politicians in the names of streets, squares and schools in various municipalities in the state and, in particular, in Três Lagoas - the scale of this research.

2.2 The land structure of the Bolsão in Mato Grosso do Sul

The consequences of the historical occupation of the Midwest and the creation of the state of Mato Grosso do Sul, together with the old policy of hegemony of power between landowning families, materialized in the concentration of land ownership in the countryside of Mato Grosso do Sul. The fact is that throughout the state of Mato Grosso do Sul there are extensive land holdings under the condition of value reserve - as shown in table 3 of the land structure - especially in the regions where economic activity was mainly concentrated on livestock.

In this sense, extensive livestock farming still predominates in the Bolsão Rural Territory, even with the introduction of other activities, such as eucalyptus cultivation.

The Bolsão Rural Territory was part of the public policy of implementing Rural Territories, via the National Program for the Sustainable Development of Rural Territories (PRONAT), which faced a number of challenges, above all due to the impasses related to the very complexity of territorial formation, expressed in the following dimensions: environmental, socio-cultural, economic and political. In addition, the historical predominance of sectoral policies has remained rooted in the mentality of the people who make up the political spaces, making the conceptions of totality that the concept of territory demands disjointed.

Recognizing these difficulties, state policies to consolidate the Territories have emerged as tools to make the Rural Territories and Citizenship Territories effective. Thus:

> These are attempts at new types of policies that advocate a territorial basis for development policies, involving the mobilization and demands of various segments of society. As a result of identifying these difficulties as challenges, government policies have emerged in recent times (starting with the Lula government), with the aim of implementing processes for Citizenship Territories and Rural Territories (NARDOQUE; ALMEIDA, 2015, p. 3).

The Bolsão Rural Territory was consolidated[24] through the "Territorial Development Extension Center (NEDET)" project, coordinated by Professor Dr. Sedeval Nardoque. The project in question was funded by the former Ministry of Agrarian Development (MDA), through a partnership with the National Council for Scientific and Technological Development (CNPq), and provided policy advice for the National Program for the Sustainable Development of Rural Territories (PRONAT) (MELO, 2021).

The Bolsão Rural Territory is located in the eastern region of Mato Grosso do Sul (figure 3) and is made up of eight municipalities: Água Clara, Aparecida do Taboado, Cassilândia, Chapadão do Sul, Inocência, Paranaíba, Selvíria and Três Lagoas. The Territory was created by the Ministry of Agrarian Development in 2013 and implemented in 2014, based on the strategy of promoting rural development by "decentralizing" political decisions (MELO; SILVA, 2016).

Allied to the creation process, the Bolsão Rural Territorial Development Extension Center was set up, linked to the Territorial Studies Laboratory of the Federal University of Mato Grosso do Sul (UFMS) (NARDOQUE; ALMEIDA, 2015).

Map 3 - Rural Territory of Bolsão/MS, 2021

---

[24] The project selected was "Implementation and Maintenance of the Extension Center in Territorial Development of the Rural Territory of Bolsão (MS)", through CNPq/MDA/SPM-PR Public Call No. 11/2014.

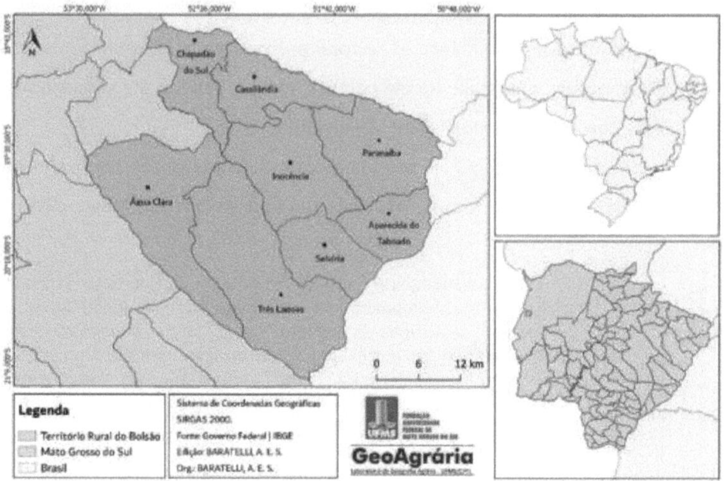

The Bolsão Rural Territory covers 45,929.9 km² and has an estimated population of 233,297 inhabitants, according to the latest IBGE Demographic Census of 2010. The municipality of Três Lagoas is the most populous, with around 119,465 inhabitants, followed by Paranaíba, with 42,010 inhabitants. In terms of land area, Três Lagoas (10,206.949 km²) is the largest municipality, followed by Água Clara (7,809.211 km²), according to Table 4.

Historically, the Bolsão had beef cattle farming as its main economic activity. There are also other activities, such as industry, commerce, family farming and the introduction of dynamic agro-industries, as shown in Table 4. As far as rural activities are concerned, since 2006 there has been an intensification of eucalyptus monoculture, along with the territorialization of the cellulose-paper complex, as the region's main economic activity.

Table 4 - Bolsão Territory: general data, 2021

Municipality	Area (km²)	Population	Main economic activities
Água Clara	7.809,211	15.257	Livestock, eucalyptus monocultures
Ap. do Taboado	2.750,150	25.431	Industry, Livestock, sugarcane/ethanol
Cassilândia	3.649,725	21.876	Livestock, agriculture and trade
Chapadão do Sul	3.248,120	24.559	Agriculture, Livestock

Innocence	5.776,028	7.625	Livestock, trade
Paranaíba	5.402,652	42.010	Livestock, commerce, agriculture
Selvíria	3.258,326	6.515	Livestock, industry, commerce
Three Lagoons	10.206,949	119.465	Industry, livestock and eucalyptus/cellulose
TOTAL	45.929,9	15.257	

Source: IBGE (2021); Nardoque, (2016). **Organized** by the author.

The large properties used to raise cattle in the Bolsão were formed through the appropriation of land by the Garcia Leal family and their members since the 19th century, with the opening of new properties, forming large possessions from Paranaíba to Três Lagoas. The agricultural and cattle raising activity expanded because it was part of the route to other provinces, such as Mato Grosso, Minas Gerais and São Paulo (LEONARDO, *et al.*, 2021).

Cattle farming continued to expand in the Bolsão/MS and reached its peak in 2005, with a total of 3,881,990 head, according to Table 5. However, this number fell in the following years, as the first company in the eucalyptus-cellulose complex, Suzano S/A (launched at the time as the VCP-IP complex), began the process of setting up in 2006, and Eldorado Brasil followed in 2010, both in Três Lagoas. In 2019, the number of heads fell to 2,585,334, a drop of around 33.39% compared to 2005.

Table 5 - Bolsão Rural Territory: cattle herd (1980-2019)

Year	Herd size
1980	1.541.739
1985	2.309.035
1990	2.953.770
1995	3.727.302
2000	3.678.412
2005	3.881.990
2010	3.332.252
2015	2.907.956
2019	2.585.334

Source: IBGE - Municipal Livestock Survey. **Org**. Leonardo *et al.*, 2021.

The municipalities that stood out most in terms of livestock were Três Lagoas, Paranaíba and Água Clara. Três Lagoas had 938,008 head of cattle in 2005 and in 2019 registered 540,685 head, a reduction of around 42.35%. Paranaíba reached its peak in 1985, with a herd of 638,302 head and, in 2019, it reached 481,942 head, a reduction of 24.49%. Água Clara also reached its highest number of heads in 2005, with 803,606 and fell to 451,887 in 2019, a percentage of around 43.77% (LEONARDO, *et al.*, 2021).

From the point of view of the land structure, the Bolsão Rural Territory followed the same characteristics as Mato Grosso do Sul, with concentrated land and unproductive rural properties. As investigated by Oliveira (2008), 23.67% of properties in Mato Grosso do Sul are unproductive. Três Lagoas and Paranaíba - both in the Bolsão - have the highest number of unproductive properties in the state, respectively: 149 unproductive properties in Três Lagoas and 129 in Paranaíba (LEONARDO, 2020).

In addition, the municipality of Três Lagoas - which has most of the state's unproductive properties - also has a high level of absenteeism, considering that around 81% of landowners do not live in the municipality (NARDOQUE, 2017). In other words, only 19.25% live in Três Lagoas, 4.81% live in other cities in Mato Grosso do Sul and 67.75% do not live in the state. Other municipalities in the rural Territory of Bolsão also stand out in terms of absenteeism rates, such as Selvíria, where only 0.52% - one owner - lives in the municipality and 32.1% are in other cities in MS and 55.79% do not live in MS. Furthermore, only in Aparecida do Taboado and Paranaíba do 50% of the owners live in the municipality, respectively: 51.45% in Aparecida do Taboado and 52.13% in Paranaíba, according to the data in table 6.

Table 6 - Bolsão Rural Territory: residence of landowners - 2010

Municipalities	Residence							Total
	Municipal headquarters		Others from MS		Outside MS		Nonexistent	
Água Clara	30	5,45%	139	25,27%	356	64.73%	25	550
Ap. do Taboado	71	51,45%	3	2,17%	41	29,71%	23	138
Cassilândia	65	26,42%	27	10,97%	140	56,91%	14	246
Chapadão do Sul	35	22,01%	52	32,7%	65	40,88%	7	159
Innocence	29	7,97%	75	20,6%	218	59,89%	42	364
Paranaíba	171	52,13%	19	5,79%	113	34,45%	25	328
Selvíria	1	0,52%	61	32,1%	106	55,79%	22	190
Three Lagoons	148	19,25%	37	4,81%	521	67,75%	63	769

Source: INCRA, 2010. **Organized** by KUDLAVICZ, M; NARDOQUE, S. (2016).

The high rate of absenteeism reinforces the idea that land in the Bolsão is strategically used as a store of value. According to the land structure data (table 7), the presence of large properties is revealed, since establishments over 1,000 hectares represent only 13.95% of the total number and occupy 2,822,213 hectares, concentrating 71.57% of the region's land. The properties of up to 50 hectares and 50 to 100 hectares combined have 2,771 establishments, or 44.33% of the total number of rural establishments and occupy 2.02% of the total area, according to the data shown in table 7.

Table 7 - Bolsão Rural Territory: Land Structure 2017

Total area class (ha)	Agricultural Census 2017				
	No. of establishments	%	Area (ha)	%	Average area
0 to minus 50	2.101	33,61%	40.681	1,03%	19,36
50 to less than 100	670	10,72%	39.219	0,99%	58,54
100 to less than 200	683	10,92%	100.407	2,55%	147,01
200 to less than 500	1.128	18,04%	369.075	9,36%	327,19
500 to less than 1000	798	12,76%	571.960	14,50%	716,74
Above 1000	872	13,95%	2.822.213	71,57%	3.236,48
Total	**6.252**	**100,00%**	**3.943.555**	**100,00%**	

Source: IBGE - Agricultural Census, 2017. **Organized** by the author.

The historical presence of cattle ranching in the Bolsão contributed to the concentrated land structure, since cattle ranching served as a strategy to support the maintenance of large properties. In Brazil, extensive cattle ranching historically covered up idle land, protecting it from land reform processes.

However, the presence of extensive cattle ranching alone does not explain the process of setting up the concentrated land structure in the Bolsão. The constitution of large rural properties is based on the recent past of land invasion, in which colonizing families decimated indigenous peoples in order to appropriate the territory, according to the literature already cited on the subject. In addition, land grabbing and the transfer/donation of state titles to regional oligarchies also contributed to land concentration, as shown in section 3.1 of this paper.

The Bolsão has 572,974 hectares of eucalyptus distributed among its municipalities, around 49.06% of the total area planted in the state of Mato Grosso do Sul, given that the state has 1,124,969 hectares planted. The municipalities with the largest plantations in the Bolsão/MS are, respectively: Três Lagoas (263,690), Água Clara (131,942) and Selvíria (87,321), as shown in Table 8.

Table 8 - Rural Territory of Bolsão/MS: Area planted with eucalyptus in hectares (2013-2019)

Municipalities	2013	2014	2015	2016	2017	2018	2019
Água Clara	95.000	118.000	120.000	125.000	126.000	128.000	131.942

Aparecida do Taboado	15.000	18.000	18.200	19.100	26.847	22.000	22.554
Cassilândia	1.200	4.000	2.500	3.000	2.500	2.200	2.423
Chapadão do Sul	1.500	3.000	3.200	3.000	3.500	3.800	3.659
Innocence	23.000	40.000	43.500	45.800	58.000	54.000	50.172
Paranaíba	5.000	6.450	7.048	9.000	10.000	10.800	11.213
Selvíria	46.000	95.000	74.350	81.500	110.000	88.000	87.321
Three Lagoons	140.000	200.000	217.600	230.000	245.000	263.000	263.690

Source: IBGE - Production of Vegetable Extraction and Forestry **Organized** by the author.

Map 4, of land use and occupation in the rural territory of Bolsão/MS, highlights the concentrated land structure of the rural territory of Bolsão and the significant presence of economic activities focused on eucalyptus and/or extensive livestock farming. In addition to these activities, there are crops for growing food, but even though there is resistance in peasant production, between 2000 and 2014 this activity fell by 0.02%. Over a 14-year period, eucalyptus monocultures grew in area by 6.47%. Capitalist agriculture (soybeans, cotton, corn, etc.) remained the predominant activity, despite a 4.64% decrease in area.

Map 4 - Land **use** and occupation in the Rural Territory of Bolsão/MS (2000-2014)

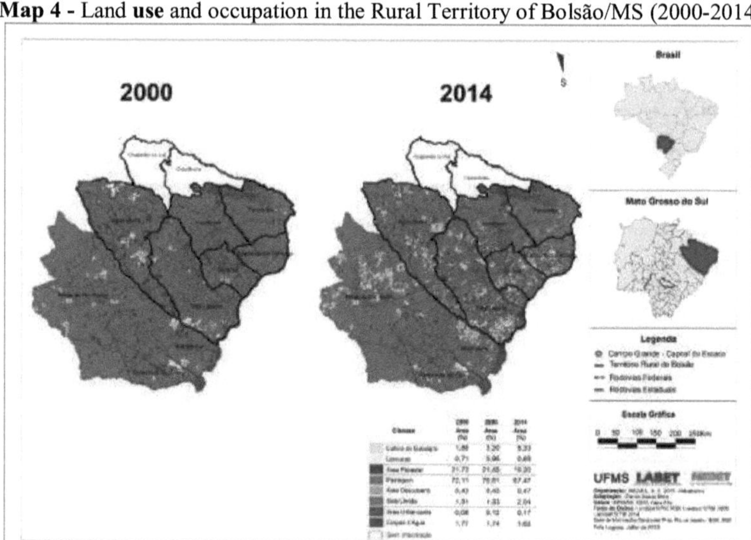

Source: São Miguel (2015), adapted.

The presence of eucalyptus monoculture plantations has further concentrated the land ownership structure in the Bolsão/MS and the activity has led to an increase in land prices. Prices for buying and renting properties have risen exorbitantly, because the demand for land to expand the monoculture has grown. This led to land (re)concentration (BARATELLI, 2019).

According to the data collected by Melo (2021), two municipalities in this territory, such as Três Lagoas and Chapadão do Sul, have seen steep rises in land prices. According to the author, in Três Lagoas, the average price per hectare went from R$3,279.00 in 2003 to R$11,250.00 in 2017, an increase of around 243.09% over a 14-year period. The author relates the rise in land prices to the expansion of eucalyptus monocultures and the growing demand for leasing and buying rural land to grow them.

The municipality of Chapadão do Sul is experiencing similar dynamics to Três Lagoas, but with even higher prices. Melo (2021) points out that the average price per hectare in the municipality went from R$9,903.00 in 2003 to R$37,500.00 in 2017. The author argues that the high prices between 2003 and 2008 were part of the advance of soybean cultivation in the municipality and that now, with the territorialization of the IACO Agrícola sugarcane mill, prices per hectare have skyrocketed.

An important dynamic has been the change in strategy of some properties that historically used to raise cattle and were leased out for eucalyptus plantations. The transfer of land to another economic activity has turned latifundia with low productivity (some of which have been labeled unproductive) into productive ones, thus making agrarian reform more difficult. That said, the increase in land prices makes Agrarian Reform unfeasible, since INCRA (the National Institute for Colonization and Agrarian Reform) does not have enough money (as already shown in topic 2.2) to buy properties at market price - especially in the current Bolsonaro government, in which Agrarian Reform and family farming policies are being dismantled (BARATELLI, 2019).

Food crops, which are the basis of Brazilians' daily diet, continue to decline in the region. Rice represents a significant drop, as there are no records of its cultivation in 2019. Beans fell by 55.88%, from 2,407 hectares in 1990 to 1,062 hectares in 2019. Cassava fell by 47.31%, from 1,860 hectares in 1990 to 980 hectares in 2019, according to the data in Table 9.

In this sense, staple food crops such as rice, beans and cassava have decreased significantly over time, from 15,367 hectares in 1990 to 6,310 hectares in the 2000s, falling to 1,970 hectares in 2010 and finally in 2019, with a slight increase to 2,042 hectares.

Table 9 - Bolsão Rural Territory: area planted in hectares with cotton, rice, beans, manioc and corn (1990-2019)

Year of cultivation	Cotton	Rice	Beans	Cassava	Corn
1990	372	11.100	2.407	1.860	24.380
1995	477	8.257	220	1.300	50.672
2000	8.154	4.770	830	710	34.817
2005	17.500	63	637	750	22.828
2010	10.256	20	1.590	360	34.752
2015	6.987	110	750	360	43.610
2019	13.772	-	1.062	980	48.980

Source: IBGE - Municipal Agricultural Production. **Org**. Leonardo *et al*. (2021).

Cotton and corn, in contrast to other crops, have grown exponentially. The area under cotton grew from 372 hectares in 1990 to 13,772 hectares in 2019, an increase of around 3,602% in the area planted and 15,172% in the quantity produced. Corn grew from 24,380 hectares in 1990 to 48,980 hectares in 2019, an increase of around 100%, and the quantity produced also grew by 376% (LEONARDO, *et al*., 2021). According to 2020 data from IBGE Sidra, corn cultivation reached 49,986 hectares, of which 34,000 are located in the municipality of Chapadão do Sul.

The production of basic foodstuffs that go straight to the tables of Brazilian workers, such as rice, beans and manioc, are grown by family farmers. Despite the concentrated land ownership structure; the strong presence of unproductive latifundia and, in the most recent period, the introduction of productive latifundia, the peasants of Bolsão continue to exercise their farming practices based on the stubbornness of those who (r)exist the imposition of the capitalist model of agriculture (KUDLAVICZ, 2017).

Peasant family farming in the Bolsão, based on the principles of growing quality food, has included agroecological management in its practices, i.e. clean production without pesticides (poison) and chemical fertilizers. Adherence to this type of management is linked to the presence of the UFMS in the rural settlements, especially in the 20 de Março Settlement, located in the municipality of Três Lagoas, which is part of the research project to set up the Center for Agroecological Studies (NEA) .[25]

In line with the principles of peasant family farming and the struggle for the social reproduction of settled families to occur with autonomy in production and marketing, avoiding middlemen, NEA has created Agroecological Fair projects at UFMS/Campus of Três Lagoas and in condominiums in the city of Três Lagoas, which currently take place *online* due to the covid-19 pandemic.

---

[25] The project was approved by the National Council for Scientific and Technological Development (CNPq) through call MCTIC/MAPA/MEC/SEAD - Casa Civil/CNPq No. 21/2016.

CHAPTER 3: LAND CONCENTRATION AS A PRODUCT AND CONDITION OF POWER RELATIONS IN THE MUNICIPALITY OF TRÊS LAGOAS/MS

The hegemonic power relations in the municipality of Três Lagoas remain based on the interests of those who own the land. The practice in question is rooted in the historical legacies of the Brazilian process of territorial formation with invasion of the territory of the original peoples, of which the municipality is a part.

It is, therefore, a heritage whose core is struggle, based on a great deal of violence against the original peoples with the aim of concentrating land and forming latifundia, as discussed in the previous chapter

The rentier agrarian oligarchies controlled the process of forming the municipality by donating/buying land and maintaining the inequalities between land owners and the dispossessed. Therefore, the concentration of land was formed in the territory and kept as a store of value, for the purposes of speculation and the realization of land rent. The practices of land concentration and rent extraction take place both in the countryside and in the city, because the owners of these fractions of the territory are often the same.

The rentier elite also seeks to make itself present in public power, acting in favor of their class benefits, based on the practice of using land as a commodity capable of generating financial income, increasing their wealth. This historical culture of using the ideological apparatus of the state to guide the debate on capitalist development has condemned the municipality of Três Lagoas to being a territory of concentrated wealth and great social inequality, currently represented by two extremes: it is considered the "national pulp capital", with international reach, and at the same time, it is a place where a significant number of poor people live, experiencing social ills.

The analysis of the land-capital contradiction was made possible by understanding the presence of an elite rooted in the production of the territory of Tres-Lago. The dynamic of seeking to earn income and profit, to the same extent, is evidenced by the way in which traditional families, in the municipality's politics and economy, have followed the neoliberal wave to combine their interests as landowners with the business dynamics of the expansion of eucalyptus monocultures, thus increasing their percentages of earning land income and profit.

The contrast in the municipality's capitalist territorial production - which results in one-way wealth drainage, in other words, based on the enrichment of private groups - is laid bare in social ills in part of the city, represented in quantitative data from SEBRAE, SEMAGRO and the Participatory Master Plan, as well as in the images from the fieldwork carried out in the neighborhoods selected for the research.

3.1 Territory as a concept in question

Geography has traditionally defined certain categories of analysis for understanding and studying geographical space. These categories allow space to be analyzed from different perspectives: territory, place, region and landscape.

Geographical researchers have taken it upon themselves to hold discussions that contribute to understanding the relationship between society and nature in geographical space. In this sense, the complexity of social relations must be interpreted from the point of view of the totality of society in convergence and coexistence with nature (SANTOS, 1985).

The need to define categories for analyzing geographical space is based on the principle that social relations, the symbiotic relationship between society and nature, defines the need for the category used. In this way, even if the analysis is carried out on the same portion of space, it is possible for the categories to be used according to the urgencies determined by society.

It is in this sense that research in the field of geography needs to define the category to be studied, since this theoretical-methodological exercise allows the researcher to direct their gaze towards the issues to be highlighted. In the case of this work, the category adopted as a form of analysis was territory.

Territory has a wide range of conceptual definitions, since some authors interpret it differently. Renowned researchers, such as Fernandes (2013), define territory as multiple, with a different order depending on the social relationship played out in it. However, it is not our place to dwell on this debate. We will therefore focus our efforts on the concept of territory we have adopted, which is that of the single capitalist territory (Oliveira, 2007).

Although there are various theoretical disagreements about the definition of the concept of territory, the concept of territory as a category makes it possible to understand power relations, disputes and the political dimension of territory. This initial definition is part of the studies by Raffestin (1993), who defined power as an interpretative key for the concept of territory, as follows: "in every relationship, power circulates, which is neither possessed nor acquired, but simply exercised" (RAFFESTIN, 1993, p.7).

According to Moraes (2005), the use of the concept of territory in geographical analysis prevents the reproduction of traditional geography, in which territory was a simplistic definition of any portion of an area that was occupied. Thus, if the concept is to be used while preserving its political dimension, the relationships played out by the social groups occupying the territory must be taken into account. This interpretative key, which places society at the heart of the concept's definition, prevents it from being used solely on the basis of natural reality:

> A few words are in order about the concept of territory itself and its use in detriment to other more usual concepts in geographical literature, such as habitat, region or area. Its choice is based on the fact that social use is its defining element. In other words, it is appropriation itself that qualifies a piece of land as a territory. Therefore, this concept is impossible to formulate without recourse to a social group that occupies and exploits that space, the territory - in this sense - not existing as a purely natural reality [...] (ibidem, p.45).

Moraes (2005) also argues that using the concept of territory allows us to understand geographical space based on the dynamics of the society-space relationship. In this way, the naturalistic conception of territory is banished. The geographical objective, in this case the territory, therefore needs to be analyzed from a social perspective that considers its constant and dialectical movement in the formation of processes. Thus:

> [...] this concept thus has two advantages: it prevents any return to naturalistic concepts (which are so characteristic of traditional geography) and it points to a social vision of the geographical object, no longer seen as a place (the landscape or the surface of the earth), but directly as the society-space relationship itself. What's more, equated as a moving entity - formation - it also rescues the dialectical unity between form and process, which is vital for the geographical perspective that is being sought (MORAES, 2005, p. 45).

Other authors, such as Calabi and Indovina (1973) also agree with the social dimension of territory, since it is occupied and produced by society. According to these authors, territory is the product resulting from the process of social domination, inscribed in a dynamic of political and power relations, subject to capitalist logics:

> Capitalist relations of production tend to expand and encompass the whole of society; it is these relations and the development of productive forces that give the territory its specific configuration (CALABI; INDOVINA, 1973, p. 2).

Despite the determination of territory as a concept that includes an understanding of the political and power relations played out in society, there are theoretical understandings that territory is unrelated to the dispute between classes. Abramovay (1998), for example, defines territory as having multidimensional and integrating characteristics, capable of bringing together different subjects based on their belonging to the space. For this author, what defines the participation of subjects in the production of territory is the link with the space, regardless of the social class to which they belong.

However, it is essential to include the class debate in order to explain territory in the reality of the capitalist system, which is essentially based on the division of social classes. It is therefore necessary and possible to include a class perspective in the formation of territory. We therefore agree with Oliveira (2007) when he understands territory from the perspective of contradictory social and production relations:

> Territory is thus a concrete product of the class struggle waged by society in the process of producing its existence. Capitalist society is based on three fundamental classes: the proletariat, the bourgeoisie and landowners. Thus, it is the social relations of production and the continuous/contradictory process of development of the productive forces that give the territory its specific historical configuration. Territory is therefore not a prius or an a priori, but the continuous struggle of society for the equally continuous socialization of nature. [...] It is this contradictory logic that constructs/destroys territorial formations in different parts of the world or makes fractions of the same territorial formation experience unequal processes of valorization, production and reproduction of capital, shaping regions. (OLIVEIRA, 2007, p. 3).

For the author, territory is the product of the struggle between three social classes: the proletariat, the bourgeoisie and landowners. This dispute is waged and inherent in the development of the capitalist system, which contradictorily dominates, destroys, constructs and fragments the territory, a product of and produced within the capitalist system and social struggles.

The disputed territory is also produced by resistance, as in the case of social movements that occupy the territory and fight for its consolidation. However, social movements don't have a complete monopoly on the territory, but sometimes they manage to obtain fractions of it for the reproduction of life. Territory is therefore the product of a constant dispute between social classes, even if there are hegemonic classes that sometimes exercise greater control, they do not have total control, as Fabrini (2011) explains:

> But the territory is a disputed space, because peasants do not exercise total and complete domination/control/power over the space. If peasants don't have total control over the territory, capital doesn't have a complete monopoly either, because while there are relations of domination and exploitation, there is also solidarity, community spirit, mutual aid and subsistence production. In this sense, the territory is a disputed space in which one class is hegemonic, but does not have total control. (FABRINI, 2011, p. 103).

The disputes between antagonistic classes for control of the territory narrate and constitute the history of society. Moraes (2005) therefore defines that social history is demarcated on the earth's space, imprinting it through the process of territorial formation. Every social relationship exerts the movement of territorial formation:

> Territorial formation is therefore one of the defining elements of particularity (now thought of, for example, on the scale of "national peculiarities"). In short, historical development takes place on and with land space and, in this sense, every social formation is also territorial, as it necessarily specializes. (MORAES, 2005, 47).

Adopting the concept of territorial formation to explain the processes that took place in the formation of Mato Grosso do Sul and the municipality of Três Lagoas allows us to understand the naked presence of social relations of power and dispute in the production of the territory. These relations were played out mainly by rentier landowners, who are sometimes confused with industrial capitalists and part of the state, and also with the territorialization of capitalist agricultural companies. However, despite the fact that the hegemonic classes maintain control of fractions of the territory, it is worth remembering that the dispute is dialectical and constant, driven above all by the struggle between antagonistic classes, which also drives the processes of resistance by peasants and the working class.

3.2 The invasion of the territory and the violence to maintain it: from sertanejos to colonels

The history of the legal formation of the municipality of Três Lagoas went through several stages. At first, Três Lagoas was considered a district of the municipality of Santana do Paranaíba, via state law no. 656, of 12-06-1914. In 1915, Três Lagoas was elevated to the status of a town in the district, by state law no. 706, of 15-06-1915, which separated it from Santana do Paranaíba. It was then, in 1920, due to a change in state resolution no. 820, of 19-10-1920, that Três Lagoas became a city and the municipal seat.

However, the occupation of the area corresponding to Três Lagoas began long before its legal formation. The history of the municipality of Três Lagoas is based on the expeditions of the bandeirantes to clear territories, based on greed for land and violence against indigenous communities, such as the Cayapó and Ofaié who inhabited the Três Lagoas region. According to Mendonça (1991), the first to occupy the land were the bandeirante Antonio Pires de Campos and the sertanista Joaquim Francisco Lopes.

Antonio Pires de Campos began his expeditions through the backlands of Mato Grosso between 1722 and 1726. Joaquim Francisco Lopes started in the following century, from 1828 to 1839, coming from the Triângulo Mineiro with 11 other people to occupy land in the Monte Alto region. It was here that his entourage met the brothers José Garcia Leal and Januário Garcia Leal, who had settled near the Rio Grande and Paranaíba.

Joaquim Francisco Lopes' expeditions in search of land continued and were joined by the Garcia Leal brothers. As the group advanced through Mato Grosso, they expanded their properties, the aim of which was to locate fields suitable for cattle breeding. Due to the presence of indigenous communities, such as the Cayapó and Ofaié, the conquest of the land was based on conflicts with indigenous resistance and genocide (MENDONÇA, 1991). Thus:

> And the spirit of adventure united these men who were not limited
> to the excitement of the race for prey, but also driven by greed for
> land: land... more and more land! (MENDONÇA, 1991, p. 61).

The families of Francisco Lopes and Garcia Leal continued to conquer land in Mato Grosso, especially in the southern region of the state. According to Mendonça (1991), the first to start settling near the location that would become Três Lagoas was Protázio Garcia Leal, grandson of Januário Garcia Leal. The conqueror in question carried with him his backwoods past and the heritage of being part of the family of "the main man of the Paranaíba backwoods", his grand uncle José Garcia Leal.

Other groups of sertanejos and cattle ranchers also made inroads into the region's land, such as the families of Antonio Trajano dos Santos, Costa Lima and Queiroz. These families set up their estates and became the driving force behind the formation of the municipality of Três Lagoas.

The sertanista past allowed these families to take possession of several large estates, as well as becoming the organizers of the municipalities that sprang up in their invasion territory. According to Mendonça (1991), for these families, land was treated as a precious commodity and the armed struggle to preserve their possessions took place surrounded by unbridled violence, be it against indigenous communities and/or anyone else who tried to take it. Thus, the traditional adventurous sertanistas became colonels[26] . As shown by Mendonça (1991):

> They were sertanejos, they became sertanistas; they prospered as ranchers, they became colonels. These ranchers' greed for land and power, and the weakness of the central government, marked by its absence, forged decades of mischief.
> This was the moment when the profile of those who would be the organizers declined, and it was through the concentration of land that they organized the space. Here, local bossism was formed at the same time as land grabbing, integrating itself, from the outset, into coronelismo, just as the latter was reaching its full expansion in the First Republic (p. 65).

The mass movement of people to the Três Lagoas region was due to the arrival of workers from other regions to build the railroad. It was at Km 33 of the Noroeste, at the first railway station in the south of Mato Grosso, that Três Lagoas was founded.

However, the way in which the migrants and immigrants took possession of the land was selective. New characters were added to the traditional planners of the territory, characterized as "openers of the hinterland": doctors, lawyers, journalists and

---

[26] Coronelismo is a system that originated in the old republic and manifests itself in a "compromise", an "exchange of benefits" between the political leader and the state government, with the latter meeting the interests and demands of the rural electorate (FAORO, 1958, p. 749).

pharmacists from São Paulo and Rio de Janeiro. European and Asian traders also immigrated to the region (MENDONÇA, 1991).

Immigrants and migrants who were part of the hegemonic groups dominated huge areas in Três Lagoas and became part of the municipality's coronelista rulers. On the other hand, the "peons" who worked on the construction of the railroad were denied the possibility of occupying land to reproduce their lives. According to Mendonça, "in addition to land, they were deprived of decent housing, food and work" (1991, p. 67).

The configuration of these new landowners in the municipality led to intense and violent disputes with the old group of local landowners, the essence of the conflict being competition for land and local power. According to Mendonça (1991), the effervescence of the differences between the "new" and "old" landowners brought banditry to the fore as an alternative way of fighting back. According to Mendonça:

> Banditry became the instrument of the colonels and the tradition of violence and impetuosity was absorbed by the newcomers. The conflict between the 'old' and the 'new' owners of the land was, at first, over land, but was later revealed to be over command, over the power to control the social and political organization, over ranks, positions and influence. These clashes and attacks generated within power legitimized violence, institutionalizing it as a form of order (MENDONÇA, 1991, p. 68).

The issue of disputes between power holders in Três Lagoas, whether over land and/or local power, can be understood through the concept of coronelismo proposed by Leal (1997), who characterizes such practices as:

> [...] above all, it was a compromise, an exchange of profits between the public authorities and the local bosses, especially the landlords. It is therefore impossible to understand the phenomenon without reference to our agrarian structure, which provides the basis for the manifestations of private power that are still so visible in the interior of Brazil (p. 40).

The presence of the coronelista dispute in the early days of the municipality's territorial formation, between old landowners, divided the town into two "clans". According to Mendonça (1991), the first clan was made up of the families considered to be the first owners of the land: the Queiroz, the Garcias, the da Costa, the Lopes, the Francos and the Souzas. The other clan was made up of business owners, mostly immigrants. These two clans were divided into landowners, merchants and representatives of the local government (MENDONÇA, 1991).

While some of the immigrants and migrants were conquering the territory as part of local bossism, the poor migrants, especially those who worked in the countryside, lived the reality of labor poverty. They sold their labor in any sector, whether as farm hands, cattle drovers, foremen, farmers, newsagents, etc. They lived in sheds, farms or ranches and reproduced life amidst the violence of executions contracted by local officials. It was no different in the city. Most of the population were railroad workers, treated like "beasts of burden". The work was uninterrupted for weeks on end and the living conditions were unhealthy, similar to corrals (MENDONÇA, 1991).

In addition to the fact that these rural and urban workers lived through social problems, in unhealthy working and survival conditions, the urban planning in Tres-Lago deprived them of access to land, since in order to occupy the municipality's land they had to obtain a definitive title, granted by the Land Section in Cuiabá, and pay taxes to the government (MENDONÇA, 1991).

In addition to the bureaucratic processes for occupying the land, the arm of state violence was used in the concession process. The conflict between landowning families and the persecution for increasing their areas triggered long-lasting disputes, such as the dispute between the Costa Lima and Faustino Franco, who disputed 400 "leagues"[27] of the Taquarussu farm - a vacant area owned by the state - which was to be divided between Bento Ribeiro and the English and American immigrants. The intensification of the conflict led to the immigrants being expelled from the area. The state's solution to the problem in question, which involved 829,376 hectares, was to distribute "quinhões" (shares), of which around 500,000 hectares were donated to the English, where they formed the farm & P.C. and Walter George Waldson. The remainder was distributed to farms such as that of Senator Vitorino, Colonel Álvaro Feijó, Costa Lima and Garcia (MENDONÇA, 1991).

The vast amount of land; the unbureaucratic process of concessions/donations; the advance of cattle breeding and the economic growth of the southern region of Mato Grosso, added to the intensification of the separatist agenda, attracted the presence of surnames from the elite of northern Mato Grosso, such as Celestino, Azeredo, de Toledo, Corrêa and Muller, intensifying the political disputes between northern and

---

[27] The term "léguas" was used as a unit of measurement for Portuguese itineraries. In terms of size, a légua can vary between 4 and 7 kilometers.

southern Mato Grosso. The group, which founded the Republican Party of Mato Grosso, also became involved in the politics of Três Lagoense (MENDONÇA, 1991).

In contrast to the party made up of names from the north of Mato Grosso, the clan of colonels from Três Lagoas also founded their own party, the aim of which was to support their political appointees and maintain control of the municipality.

> A new party was created, led by colonels Alfredo Justino, Januário Leal, Protásio Garcia, Faustino Franco, José Carlos and Antonio de Souza Queiroz. Afonso Garcia Prado, Juca Faustino and other sertanejos were also part of the new faction (MENDONÇA, 1991, p. 219).

However, even with the presence of different groups, but with similar political premises, whose essence remained in political coronelism and the appropriation of land as a store of value, the pacts were maintained. There was, therefore, the fact that both groups sought political control of the municipality through different parties, but because they were part of the landowning class, regardless of the governorship in question, they had identities. In this way, the rentier oligarchic groups vying for power ended up joining forces to prevent the poor community from occupying the land.

The selective granting of state land to the immigrant and migrant population laid bare the elitist and prejudiced face from which Três Lagoas originated. Urban land, divided into suburban plots in 1917, was granted mainly to foreigners, even though there was an intense influx of poor northeasterners fleeing the drought. Thus:

> It would have been possible to establish a significant number of migrants there, because it was precisely from this date that the flow of northeasterners plagued by drought grew. However, they weren't the ones we wanted; the press repeated the appeal for a 'foreign arm, without which any attempt will be futile'. The justification was that this was 'a new country in which the settlement of the soil', being 'a problem of great importance', needed an 'undertaking of magnitude', which would only be satisfied with foreign labor (MENDONÇA, 1991, p. 161).

Both the press and the state invested in the discourse of land donations to attract the foreign population, based mainly on the principle of the superiority of immigrants. In 1921, 25 plots of agricultural land were granted to those who could prove they were settlers and had lived in Brazil for less than a year. The size of the plots given away free of charge could not exceed 500 hectares. Brazilian migrants who were assumed to be poor were given 25 titles of up to just 10 hectares (MENDONÇA, 1991).

Even with the demands about such injustice, as protested by councilman Bruno Garcia and Olintho Mancini, in favor of distributing land to the poor, there was no agreement. In order for the poor population to continue to be dispossessed of land, the strategy was to charge high prices for it, as shown by Mendonça (1991):

> In any case, the poor would be rejected from land distribution. Resolution no. 46 of 1922 created mechanisms to expel them if they received a provisional title. One article excluded them in advance: 'The land owned by the municipality that is requested for leasing, in the suburban area of this city, from this law onwards, will be paid at the rate of $005 per m² and granted on condition that it is leased and improved. The closure, even with wire, is considered a benefit in the first year of the concession'. The free concession to foreigners was an exception, and the explicit refusal to poor natives was nothing more than a reinforcement (p. 163).

In 1937, Três Lagoas was classified by the state of Mato Grosso as "a great land vacuum", and the state therefore reserved around 3,600 hectares of land for the municipality. At first, the municipality began to grant titles of up to 100 hectares, but this was based on the selective principle of favoring landowners. In 1940, there were still "empty spaces", but the public authorities did not consider donating land to the poor, since the aim at that stage was to sell it. The commodity nature of land was identified by the mayor, Lt. Col. Manoel Pereira da Silva, who explained the problem of the low take-up of land in this way: the high price. It was then that the municipality granted 300,000 reis for suburban plots of up to 10 hectares, i.e. at a price of 30,000 reis each (MENDONÇA, 1991).

Maintaining access to land through purchase had been questioned once again by councilman Dr. Bruno Garcia, in 1940, who argued that the south of Mato Grosso would become an area of large estates and, in order to alleviate this situation, a "modest agrarian revolution" had to take place, with the donation of small plots to those who undertook to cultivate them. But the proposal was voted down (MENDONÇA, 1991).

During the establishment of town planning and the sale of plots, a number of laws were passed. One of the laws that was voted on and approved was the collection of taxes which, if neglected in three years, would result in expropriation. However, the law was not applied equally to all individuals. The possibility of negotiating taxes in arrears from 1919 to 1930 was a privilege for friends of the territorial organizers, as evidenced by the letter signed by the intendant Bruno Garcia:

> '[...] to inform my friend of the progress of the taxes on his plots of
> land [...] which amount to 343$ 200 without the respective fines that
> I can waive, imposed since 1919. I would also like to inform you
> that there are several suitors for the aforementioned plots, and I only
> hope for a resolution on your part, to redeem the taxes owed, or else
> the Intendencia will dispose of them under the law of the
> commission'. (MENDONÇA, 1991, 1968).

The characteristic interpersonal relationships present in Brazilian politics reinforce the patrimonialist nature of the Brazilian state (FAORO, 1958), since the presence of rentier interests from the beginning of territorial formation, together with representatives of local power, reinforces practices of political patronage. The formation of the municipality of Três Lagoas was no different. The donation/granting/sale of rural plots and the political benefits to representatives of public power were also repeated when it came to urban land. According to Mendonça (1991):

> In the distribution of plots throughout the period, we can see that
> their holders in the central areas were names of well-known
> authorities: councillors, delegates, judges, intendants, or prosperous
> men from commerce and livestock: Fenelon Müller, Francisco
> Garcia Leal, Siegefredo Roriz, Renato and Rômulo Carrato,
> Januário Penelli, Manoel Bazan, José Azevedo Coutinho, among
> others. Many with two or three plots and, not very often, but with
> some occurrences, the leasing of an entire block (p. 168).

Historically, the strategies of the elite in Tres-Lago to grab land revenue are essentially based on a relationship of inequality between those who own land and those who are dispossessed of it - in other words, the strategy of land monopoly. Therefore, the historical legal and political strategies of the three-lagoon elite in denying access to land to the poor is part of an authoritarian project of captive land to gain income and power. Thus:

> During the concentration of land in the hands of a minority, the task
> of exclusion was quiet in the sphere of laws and regulations, but
> fierce in the struggle for power between those who owned the land.
> Land and power, in a direct relationship - not just preserving one and
> the other, but extending them - was the motive for the struggles
> between the owners of the land. Once the distribution of the land had
> been ensured and conveniently distributed to those who had power
> over it, certain areas of society were organized: wealth, work,
> housing and leisure. The place and significance of the dispossessed
> in this community were demarcated (MENDONÇA, 1991, p. 170).

The fact is that the reality in Três Lagoas illustrates the rentier nature of the traditional and modern Brazilian elite. In Brazil, and also in the municipality of Três

Lagoas, the appropriation of land is synonymous with power. On the other hand, the concentration and control of access to land makes it possible to use land as a store of value, in which the process of speculation is the most visible aspect. The aim of the land monopoly is to enable landowners to earn rent and turn it into capital. As rent is a social tax paid by society as a whole so that this finite asset, land, can be used, the debate on Brazil's territorial formation - and the agrarian question that arose from it, expressed in the land monopoly - should be of interest to the general population. Unfortunately, this is not the case even in left-wing political parties, partly because the ideology of land as a commodity, synonymous with development and the progress of capital, has become more important than human development and social justice.

3.3 Land: as a codename for power and its effects on the municipality of Três Lagoas

In the municipality of Três Lagoas, land has always represented (and still represents) the hegemony of local power. The concept of power as something intrinsic to the control of land is part of the rentier model of capitalism that has been structured in Brazil since the mid-19th century, and is part of the legacy of the municipality's territorial formation process, as shown in section 4.2, in which traditional oligarchic families were confused with political representatives. Nowadays, some government policies of an industrial nature have been created under the guise of developing the municipality, but they have largely been taken advantage of and converted to private class interests, to the detriment of social needs.

In 1974, Brazil instituted the Second National Development Plan, the aim of which was to intensify the process of national integration and industrial development in areas with growth potential (ABREU, 2001). As part of the II PND, the areas were divided up by superintendency, thus giving rise to SUDECO, whose history is explained in section 2.2.

From the point of view of national integration, SUDECO continued the actions of the Brasil Central Foundation, which was based on integrating the Central-West region with the Southeast of the country by means of axes of penetration. These institutional policies led to the creation of important forms of transportation in the region:

> [...] it is possible to see the shape of a fan, starting from São Paulo,
> towards the interior, while the main axes in the Center-West have
> little connection with each other. They are: Belém-PA/Brasília-DF
> (BR-153); Campo Grande-MS/Cuibá-MT/Santarém-PA (BR-163);
> Cuibá-MT/Porto Velho-RO (BR-364), which opens the connection
> with the Transamazônica, through the north of Mato Grosso; Três
> Lagoas/Jataí/Aragarças/Altamira (BR-158) and BR-174 and BR-
> 242, which follows on from BR-080, among others. (ABREU, 2001,
> p. 63).

The municipality of Três Lagoas was also a priority area for investment by POLOCENTRO, which, through the Campo Grande/Três Lagoas hub, invested in transportation facilities for around 227 km, spending around Cr$196,811,000.00 (US$2.12 million) (ABREU, 2001).

Among the objectives set by SUDECO, one that received the most investment was national integration for territorial fluidity. To carry out the project in question, the municipalities of Campo Grande, Dourados, Três Lagoas and Corumbá were chosen to create industrial districts. The cities were chosen as hubs for industrialization. State-run districts were created in Campo Grande, Dourados and Corumbá, while the two districts created in Três Lagoas were run by the municipality (SOUZA, 2008).

Along with the incentive for Três Lagoas to become an industrialization hub, the Engenheiro Sousa Dias Hydroelectric Power Plant (Jupiá) was installed on the border between Mato Grosso do Sul and the state of São Paulo. The Jupiá plant began its construction process in the 1960s, under the government of Ademar Pereira de Barros, former governor of São Paulo (1947-1951 and 1963-1966), and was completed in 1974. The construction period significantly intensified the migration of workers for the construction process. During the construction period, these workers settled in the Vila Piloto neighborhood, which was built precisely for housing purposes.

Due to its location,[28] close to the state of São Paulo, Três Lagoas has also received other developments. In the late 1990s and early 2000s, the municipality was the site of the construction of a Thermoelectric Power Plant, as part of the Eletrobrás Ten-Year Plan. The period in question consisted of a "cluster of projects", with the construction of the Porto de Jupiá-Três Lagoas, on the Paraná-Paraguay Waterway , the interconnection with the Tietê-Paraná Waterway, BR-262, the Novoeste Railway

---

[28] The municipality of Três Lagoas borders the state of São Paulo. The nearest town in São Paulo is Castilho, about 20 km away.

(EF-265) and the SP 300 connection (from the Marechal Rondon Highway to SP) (SOUZA, 2008).

Investment, both in the transportation sector - for territorial fluidity - and in the industrial sector, was part of the intensification of the production of raw materials in Mato Grosso do Sul due to the ease of transport. Thus, the industrial areas were also programs for the territorialization of agro-industries, since the state had already established itself as the country's production hub (SOUZA, 2010).

In this sense, even with the presence of investments in industrial enterprises in Três Lagoas, agricultural activities, especially those centered on beef cattle exports, have remained strong. Historically, the presence of extensive livestock farming in the municipality has served as a strategy for escaping agrarian reform and maintaining large estates. Government credits have accentuated this process because, according to Teixeira (2005), POLOCENTRO funds for investment in soybean plantations have been diverted to increase the number of large estates.

In fact, public investment in the installation of instruments for territorial fluidity has attracted the territorialization of various enterprises, as recently happened with the pulp companies that considered the existence of these three modes of transport for their installation. However, the arrival of these companies has not changed the essence of the accumulation of the three-lagoon elite, which is based on the dynamics of territorial control in order to earn income. Quite the contrary, the developments in question served to contribute to this process, since they increased the value of the territory and raised land prices.

It is clear, therefore, that the rentier side of the three-lagoon elite has maintained its essence of accumulation based on land holdings, a situation that mirrors the land-capital alliance, where rentierism is at the center of economic and power relations. Thus, even with the introduction of public investments in agro-industrial activities; the advance of territorial fluidity; the "Green Revolution" - which intensified the rural exodus - the land structure remained concentrated, a prime example of the model known theoretically as conservative modernization (PIRES, RAMOS, 2019).

Três Lagoas has what is classic in Brazilian capitalism, namely the presence of the land-capital class alliance that blocks the inherent conflict between the landowning class and the industrial capitalists, through the actions of the financing and entrepreneurial state. Thus, it is possible for land to remain concentrated in favour of landowners, since the state plays the role of investor/financier for the territorialization of agro-industrial complexes, mitigating conflicts.

In addition to the actions of the state at the federal and state level through investments in companies and exemptions, the communion between landowners and capitalists - who are sometimes confused in the same class - is also possible through the actions of the municipal government. In the case of Três Lagoas, the traditional rentier groups are the supporters of "industrial development" and are close to the municipality's political power, which allows private projects to be carried out, especially those involving the concentration of land.

One example, of the many that exist in the municipality, happened to deputy mayor Hélio Morales (1997-2004) who, according to Case No. 0000815-12.2009.8.12.0021[29] , swapped areas. The then deputy mayor exchanged land registrations, in which his private land became public land and the public land, owned by the state of Mato Grosso do Sul, passed into his name.

In order to better understand the area swap, an interview was conducted with a lawyer from the city and, according to the interviewee, the deputy mayor used the state for his own benefit, since he swapped his land, which was located in a bad area with flooding, for a privileged area owned by the state of Mato Grosso do Sul:

> Hélio Morales, when he was deputy mayor, went to the town hall and simply changed it, he made a swap. The area that previously belonged to the municipality and which was passed on to the state for the construction of the JK Housing Complex, he went there and made a swap. He had a piece of land in a bad area, which flooded every time it rained, and he had this area that was in the public interest, so he simply made the exchange and he became the owner of this area, where it belonged to the state. (Interviewee A - 12/10/2021) .[30]

The case file also shows that the action taken by the former deputy mayor involved other issues, such as the presence of his son. According to the final legal documents, a proof of sale of the block was forged so that there would be no suspicions involving the deputy mayor. In this way, it was Hélio Morales' son who carried out the negotiations for the sale of the land and the construction of the houses, via funding from Caixa Econômica Federal, as evidenced by the verdict in the case:

---

[29] Recision/Resolution Process, No. 0000815-12.2009.8.12.0021, of 2009, of the Public Finance and Public Records Court.
[30] Interview conducted on October 12, 2021 with a lawyer from the municipality of Três Lagoas/MS.

Although the illegal act carried out by the former deputy mayor was investigated and the final findings led to him being found guilty of the actions, there was no conviction. And the judge ordered that the registrations of the houses sold by Hélio be unblocked, since the people who bought them acted in good faith and could not be harmed. In this way, the deputy mayor benefited, since he obtained a higher price for the land in the process of selling it, in a more valuable area that was free from flooding.

Legal tricks and the class pact have helped boost the monopoly on land, which is essentially understood as a source of power in Três Lagoas. Because of this, as well as concentrating it, the elite in Três Lagoas, in order to promote their social reproduction as a class, have no interest in the dispossessed having access to land and, therefore, in overcoming their condition of poverty.

It is in this sense, of denying access to land to the dispossessed, that we will analyze the exclusionary nature of public power when it comes to land distribution. An example of this phenomenon is the historic struggle for land at the América Rodrigues Camp in 1986.

The América Rodrigues Camp was organized by around 200 families, on June 6, 1986, who camped on the side of the MS-395 highway, in front of the Barra da Moeda farm - which already had signs of unproductivity - 30km from Três Lagoas. The group's aim was to negotiate the inclusion of these families in the National Agrarian Reform Plan (PNRA). However, the then governor of the state, Ramez Tebet, wanted to put an end to any encampment and had even banned its initial formation (FARIAS, 1997).

The order to ban the camp was carried out quickly, so the camp only lasted a day and a night. The next day, with great violence, the police and representatives of the state government broke up the camp and left the families on the highway to Três Lagoas to leave.

> According to CPT documents, there were more or less 200 police officers from the state of Mato Grosso do Sul, led by Aparício, the general director of TERRASUL (Mato Grosso do Sul Land Department) - one of the bodies responsible for making Agrarian Reform possible in the state. There were about four buses and several vehicles. With violence, the police tore down the shacks and several belongings were broken or lost. We can see how quickly the state activates its repression apparatus, showing organization and efficiency beyond imagination. (FARIAS, 1997, p. 136)

In Três Lagoas, therefore, land ownership has long been the privilege of a few, as shown in section 3.1. Oligarchic groups are also those who hold power over the state and use it as a way of defending their interests, as was done by former deputy mayor Hélio Morales and by Ramez Tebet, who as governor banned landless encampments. On the other hand, the dispossessed have even been denied the right to fight, as in the case of the América Rodrigues Camp.

The result of the benefits granted to oligarchic groups, to the detriment of fair distribution to the landless population, has created a highly concentrated land ownership structure in the municipality of Três Lagoas, which reverberates to this day. As shown in Table 10, the municipality has 906,849 hectares, of which 698,650 hectares, 77.04% of the total area, are owned and controlled by large landowners, with areas of more than 1,000 hectares representing only 15.55% of the total. On the other hand, those with up to 100 hectares represent 44.37% of all establishments and occupy 1.37% of the total area, i.e. 12,358 hectares. Areas between 100 and 500 hectares occupy 25.05% of the total area, or 75,029 hectares, and account for 8.27% of all establishments. Added to the large estates, areas of 500 to 1,000 hectares account for 120,810 hectares, 13.32% of the total area and 15.00% of the establishments.

Table 10 - Três Lagoas (MS): number of establishments and size of occupied area, in hectares - 2017

Area groups in hectares	Number of establishments		Territorial area	
	no.	%	no.	%
0 a 100	485	44,37%	12.358	1,37%
100 a 500	274	25,07%	75.029	8,27%
500 a 1.000	164	15,00%	120.810	13,32%
Above 1,000	170	15,55%	698.650	77,04%
Total	1093		906.849	

Source: IBGE; Agricultural Census (2017). **Organized** by the author (2021).

The large areas, from 500 to over 1,000 hectares, if added together, represent 90.36% of the total area of the municipality, with 819,460 hectares representing only 30.55% of the total number of establishments. This shows the large-scale presence of latifundia[31] . The areas of concentrated land focus their economic activities on extensive livestock farming, which sometimes serves as a strategy for escaping the Agrarian Reform process. In addition, the large estates also lease out their land for eucalyptus monocultures.

In this way, the concentration of land ownership and the dynamic of using land as a store of value for speculation stand out in Três Lagoas. This practice happens both in the countryside and in the city. Nowadays, rural landowners speculate on the price of land in the countryside in order to rent it out to eucalyptus plantation companies and/or incorporate part of their farms into the urban perimeter in order to create new allotments. In the city, according to the 2016 Master Plan, there are around 30,000 "urban voids" that are part of the speculation dynamic. On the other hand, the logic of launching new residential allotments is constant.

It is, therefore, a capitalism whose core is based on the practice of earning land income, based on the process of producing capital outside the productive process, exploiting land and nature and using them as a source of financial gain.

The elite's vocation for real estate speculation and the government's encouragement of new developments, which generate benefits for landowners - part of the government as well - have led the municipality to social inequalities that seem insurmountable. The point here is not to compare development between cities, but the

---

[31] The term latifúndio is used in the sense indicated in the book Dicionário da Terra, organized by Márcia Motta, 2010, p. 272-276.

lack of basic urban facilities, such as paving, sewage systems, etc., is evidence of an "abandoned" city, where development is not aimed at improving the lives of the population in the interests of the common good, and is therefore democratic.

The use of data from the countryside on land concentration and the presence of this process in the city as well, the result of historical processes related to the monopoly of land, is not an exercise in explaining the city based on relations in the countryside. On the contrary, it's about understanding the relationship between the countryside and the city as a complementary dialectic, integrated fractions that are constituted as a single, municipal territory. They are, therefore, fractions of the capitalist territory.

3.4 The core of land interests: territorialization of the monopoly, state/land/capital alliance/unity and land price increases in Três Lagoas/MS

Land ownership as a synonym for economic and political power in the municipality of Três Lagoas remains a central issue for understanding territorial formation and the land-capital pact, from the past to the present. Land concentration is still a current practice in the municipality of Três Lagoas as a process of producing capital outside the productive circuits, via the land monopoly. The land monopoly and rentier practices are the basis of the wealth of many of the municipality's traditional families.

However, the process has not been linear since the municipality was formed. There have been some changes in the present day, particularly in the surnames of the families that hold economic power in the municipality.

For example, despite not being part of the group of "old" and "new" town planners, as discussed in section 4.2, some families stand out for their social prestige in the town, identified by the population as protagonists in the town's territorial formation process. In addition, these groups have unique characteristics in terms of land monopoly, because they own land in the countryside and in the city.

In this way, we should emphasize that Martins' theory (1994), in using the concept of land-capital alliance to explain the process of class pacts (urban industrial capitalists and landowners), partly serves as a key to understanding the reality of Tres-Lago.

Partly because in the municipality of Três Lagoas, the logic of the land-capital alliance is reversed, since rural landowners have found an alternative in urban areas to increase the percentage of income they earn and also their profits. In this way, it is not a question of a land-capital class pact, but of class unity in terms of their sources of reproduction, in which rural landowners are also producers of urban space.

This unit, formed from the monopoly of land and the dynamics of urban investment, was made possible by the expansion of capitalist agriculture in the municipality. The traditional landowners saw the territorialization of eucalyptus as an opportunity to combine ways of earning income and profit.

The territorialization of the cellulose-paper complex has intensified the search for land in the municipality, since the large-scale planting of these monocultures requires the purchase and/or leasing of large areas. In addition, along with the territorialization of this monopoly comes the ideology of progress and development for all, which intensifies the expansion of the city and also increases urban land prices.

Thus, the increase in the price of land in the countryside and in the city, as shown by Baratelli (2019), has heated up the land market in the municipality, which has enabled those who have a monopoly on land to increase their percentage of income, either by leasing or selling. In this sense, the practices of de-characterizing rural land for incorporation into the urban perimeter are part of the strategy of exercising real estate speculation in plots and tracts, in the terms of Rodrigues (1988).

Today, the following families stand out due to their monopoly of land, wealth and social prestige, and are part of the municipality's elite: The Salomão family, whose essence is Syrian-Lebanese immigrants. The Thomé family, also of Syrian-Lebanese origin and, finally, the Prata Tibery family, migrants from Minas Gerais, in the Triângulo Mineiro region. These families made their fortunes and names in Três Lagoas and became (to this day) part of the main land and business-owning families in the municipality, both in the countryside and in the city.

These families are empirical evidence of the theory of the land-capital alliance, because they have a leading presence in the countryside and in the city, especially when it comes to launching new allotments and condominiums. The main fact is that the center of the interests of these selected groups is the realization of land income. In this way, the economic performance of their activities related to land trading and the production of urban space shows the rentier side of capitalist accumulation on the part of this elite in Tres-Lago.

In the theory put forward by Faoro (1958), as industrialization progressed there would be a tendency for landowners to open up their power to urban traders.

In the current case of Três Lagoas, this process has taken place through families who began their activities in commerce and/or municipal livestock farming and, over time, expanded and diversified their territorial domain. As the city became more industrialized (above all, with the presence of cellulose capital), focused above all on the interests of land for eucalyptus, these landowners began to produce the countryside and the city, through allotments and urban developments etc.

As Martins (1994) points out, the materialization of this pact, which is the land-capital alliance, only becomes possible due to the presence of the state as the central pillar in the development of capitalist relations of production. In the case of Três Lagoas, the formation of the land-capital unit receives continuous attention from the actions of the state, thus forming the unity of the state/land/capital triad.

It is through the ideology of progress and development for all that the state justifies the determining conditions for the territorialization of the cellulose-paper complex. The state thus serves to boost territorial fluidity and intensify financial and political resources for this territorialization. These combined actions are the core that gives unity to the state/land/capital triad, since it was possible to territorialize the eucalyptus-pulp monopoly without the need to make the local land monopoly unviable, on the contrary.

Faoro (1958), in defining the nuances of coronelista practices, highlights an important characteristic. According to the author, the colonel is not necessarily a political leader, nor does he need to be the sole holder of wealth. What grants him the title of "colonel" is social recognition/prestige, ideological power, as in: "It turns out that the colonel does not rule because he has wealth, but because he is recognized as having this power, in an unwritten pact." (1958, p.737).

In the municipality of Três Lagoas, the selected families were honored with the names of main streets and avenues, as shown in map 5. Other tributes have also been paid, such as the naming of hospitals and their wards, monuments, etc. In addition, the local media[32] recognizes and names these families as "traditional" in Três Lagoas.

Historically, these families, with the help of the local media, have established themselves in the history of Três Lagoas as important builders of the city's politics and capitalist development. Therefore, the private realization of the social reproduction of these groups is done materially, but also subjectively.

Figure 1: Três Lagoas (MS): streets named after "traditional" families

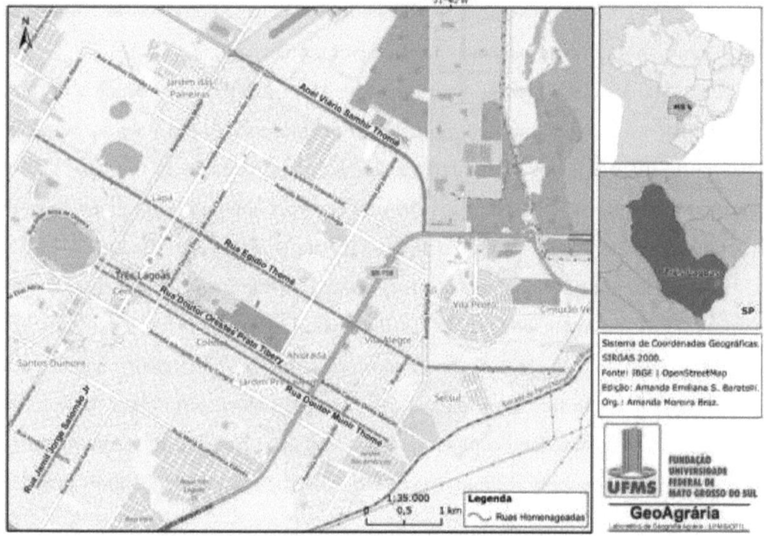

In order to achieve the proposed objective and obtain data, it was necessary to carry out methodological procedures, namely: collecting data from areas in order to draw up maps that make it possible to understand the municipality's present and past. Official information was acquired through data from the Três Lagoas Real Estate Registry Office, investigation of cases at the Mato Grosso do Sul Court of Justice (TJMS) and orientation visits to INCRA-MS. In addition, data was collected from the Federal Revenue *website*, with the aim of acquiring information on the companies

---

[32] Available at <https://www.rcn67.com.br/jpnews/tres-lagoas/se-as-ruas-narrassem-as-proprias-historias/129381/> Accessed on: January 26, 2022.

owned by the families analyzed. Finally, interviews were conducted with a former SANSUL/MS employee, a local lawyer and residents of the selected neighborhoods.

3.4.1 The Thomé family: in the city, in the countryside and in land income

The Thomé family is part of the chain of Syrian-Lebanese immigrants who came to the region during the period of land occupation and the start of local commerce. The forerunner of the family name was Egídio Thomé and he consolidated his position as part of the local elite by founding and chairing the Três Lagoas Commercial and Industrial Association (ACITL) in 1926[33] . The aim of the association was to defend the interests of local businessmen by uniting the commercial, industrial, cattle breeding and farming sectors. The family continued to be part of the traditional elite of Três Lagoas through their children.

The Thomé family's traditional activities have historically focused on commercial activities. According to the data collected by the Internal Revenue Service, the family has around 10 (ten) CNPJs, as shown in Table 11. The sum of the share capital declared by the family's companies comes to a total of R$8,232,000, considering that two of the companies have no declared share capital.

Table 11: Family businesses Thomé

Company name	Municipality	CNPJ	Start date	Share Capital (R$)
Supermercados Thomé Eireli	Three Lagoons	33.768.854/0002-52	20/02/2015	3.500.000
Supermercados Thomé Eireli	Three Lagoons	33.768.854/0001-71	19/06/1990	3.500.000
Lotérica Thomé Ltda.	Three Lagoons	19.468.094/0001-56	06/01/2014	310.000
Nova California Empreendimentos Imobiliários Ltda	Three Lagoons	18.976.318/0001-78	30/09/2013	42.000
Magid Thomé Filho - Eireli	Three Lagoons	08.719.087/0002-44	21/10/2010	100.000
Magid Thomé Filho - Eireli	Three Lagoons	08.719.087/0001-63	19/03/2007	100.000
Arrepio Producoes Musicais Ltda	Three Lagoons	27.479.303/0001-95	06/04/2017	10.000
3 Irmaos Holding Ltda	Three Lagoons	39.939.316/0001-89	27/11/2020	670.000
Águas Floresta Ltda	Three Lagoons	03.287.527/0001-37	19/07/1999	00

---

[33] Available at <https://www.acitreslagoas.com.br/nossa-historia> Accessed on: January 26, 2022.

| 3 Brothers Sand Harbor | Three Lagoons | 08.719.087/0001-63 | 19/03/2007 | 00 |

Total: 8,232,000

Source: Receita Federal, 2021.

In the service sector, the Thomé family's most notable business in the municipality is Supermercados Thomé Eireli, shown in image 1, with two stores. One of the stores is located in Parque das Mangueiras and the other in the Santa Luzia neighborhood, both at the extreme ends of the municipality.

Image 1: Supermercado Thomé commercial establishment - Três Lagoas/MS

Source: Fieldwork, 2022. **Photo:** Baratelli, A. E. S., January 15, 2022.

Although the family is dedicated to ventures in the service sector, it is also active in the urban real estate market, through the real estate company Nova California Empreendimentos Imobiliários Ltda. In addition, some of the remaining cattle breeding activities are still alive in the family tradition, through the São Thomé feedlot, as shown in image 2.

Image 2: São Thomé feedlot

Source: fieldwork 2022. **Photo:** Baratelli, A. E. S., January 15, 2022.

The expressiveness of the Thomé family in economic activities in the service sector does not exclude the characteristics of the elites of Tres-Lago in maintaining the practice of earning income from the land. Although the family's activities are centered on urban areas and the creation of new allotments, the practice of renting land reveals the rentier accumulation nature of this family, who remain representatives of the landowners. Image 3 shows a sign placed on one of the family's properties, close to the Samir Thomé ring road, advertising an area for rent.

Image 3: Disclosure of area for rent by the Thomé family

Source: Fieldwork, 2022. **Photo:** Baratelli, A. E. S., January 15, 2022.

Marx (2017), in his discussion of the forms of land rent, the purpose of which is to earn income, pointed out that the renting of raw land presupposes that the tenant must invest part of his capital by fixing it in the land, by means of improvements, building, corrections and soil, etc. to be incorporated into the constitution of his company. This form of interest incorporated into the land by the tenant forms part of the rent paid to the landowner, but is distinct from the rent for the use of the land. Marx also points out that these investments in the land are made exclusively by the tenants.

Also according to Marx (2017), the tenant sees the opportunity to increase his income by taking advantage of the investments made by the tenant, as follows:

111

> But as soon as the lease period fixed by the contract expires - and this is one of the reasons why, with the development of capitalist production, the landowner tries to shorten the lease period as much as possible - the improvements incorporated into the soil fall into the hands of the landowner as inseparable accidents of the substance, of the soil, as his property. When he signs the new lease, the landowner adds the interest on the capital incorporated into the land to the land rent itself, regardless of whether he rents the land to the tenant who made the improvements or to someone else. In this way, his income grows [...] (MARX, 2017, p. 680).

In addition to the Thomé family's rent-seeking strategy of increasing the percentage of land income by advertising bare land for rent, differential income I will also be incorporated into the income earned by the family, because the area shown in image 4 is located close to the Samir Thomé ring road and neighbors the Três Lagoas Regional Hospital - a privileged location.

Image 4: Três Lagoas Regional Hospital

Source: fieldwork, 2022. **Photo:** Baratelli, A. E. S., January 15, 2022.

The 23,000-meter Três Lagoas Regional Hospital site was donated by the Thomé family in 2014. For the people of Três Lagoas, who yearn for public facilities, and for the local media, the family has done a great deed. As Magid Thomé Filho said in an interview on the subject, donating the land for the hospital was a way of giving back for having built up their assets in the town: "We needed to give back for everything the town has done for us"[34] . The donation of the land resulted in the naming of the Três Lagoas Regional Hospital after Magid Thomé, which was declared by the Governor of Mato Grosso do Sul, Reinaldo Azambuja (PSDB) .[35]

When construction of the hospital began, the state provided the area with all the necessary infrastructure, i.e. paving, sewage system, lighting, etc. As a result, the rental price in this part of the city rose sharply, since the area has a good location and surrounding infrastructure, earning differential rent I and II. During fieldwork, an interviewee pointed out other issues in his reading of the donation of the area, in particular the Thomé family's rent-seeking strategy:

> Magid Thomé donated an area for the construction of the Regional Hospital. In reality, he has a plot of land, and since the state is going to bring all the infrastructure to the hospital, this donation, in quotes, is an exchange, right, a strategy. (Interviewee A - 03/02/2022) .[36]

The area donated for the construction of the hospital, the area with rental signs and the São Thomé feedlot were part of the family's main rural property, called Fazenda Reflorestamento e Pecuária São Thomé Br, according to registration number 40.035, annex 6. The area totaled 446.8866 hectares, divided into 12 plots, owned by Magid Thomé's children and heirs. In addition to this property, INCRA's Land Management system was able to locate another property, named Reflore, which covers 7.9 hectares, as shown in Table 12.

Table 12: Rural properties of the Thomé Family

Properties	Area in hectare	Municipality
São Thomé Reforestation and Livestock Farm Br	243,384	Three Lagoons
Make Reflore	7,9	Three Lagoons
São Thomé Reforestation and Livestock Farm Br	42,8096	Three Lagoons
São Thomé Reforestation and Livestock Farm Br Gleba 3b	5,7115	Three Lagoons
São Thomé Reforestation and Livestock Farm Br Gleba 3a	65,2035	Three Lagoons
São Thomé Reforestation and Livestock Farm Br Gleba 4b	8,8915	Three Lagoons
São Thomé Reforestation and Livestock Farm Br Gleba 1d	2,0586	Three Lagoons
São Thomé Reforestation and Livestock Farm Br Gb 1c	6,0867	Three Lagoons

---

[34] Available at <https://www.perfilnews.com.br/doacao-precisava-retribuir-tudo-que-a-cidade-ja-fez-por-nos/> Accessed on: January 26, 2022.
[35] Available at < https://www.rcn67.com.br/jpnews/tres-lagoas/magid-thome-sera-homenageado-com-nome-de-hospital-regional/96311/> Accessed on January 27, 2022.
[36] Interview conducted on February 3, 2022 with a lawyer from the municipality of Três Lagoas/MS.

São Thomé Reforestation and Livestock Farm Br Gb 1b	21,9164	Three Lagoons
São Thomé Reforestation and Livestock Farm Br Gb 1a	37,3341	Three Lagoons
São Thomé Reforestation and Livestock Farm Br Gleba 3b	5,7115	Three Lagoons
São Thomé Reforestation and Livestock Farm Br Gleba 5	3,2329	Three Lagoons
São Thomé Reforestation and Livestock Farm Br Gleba 10c	4,5463	Three Lagoons
Total area: 454,7866		

Source: INCRA Land Management System, 2020

The strategy of dividing the farm into several plots is part of the process of incorporating the hectares into the urban perimeter, through the process of de-characterizing the[37] rural area into an urban area. Fazenda Reflorestamento e Pecuária São Thomé Br has a strategic location, since it is divided by the Samir Thomé ring road, in a location that is expanding due to the installation of the Três Lagoas Regional Hospital and also because it is the closest route to the Três Lagoas Shopping Mall, the Municipal Spa and the local airport.

In this way, the paths to consolidating possessions and acquiring goods are paved by what is most intrinsic in the economic and political relations of capitalism, so that privileges are selective to those who belong to the same class. In this case, the class in question is the landowners. However, these owners don't just limit themselves to land/real estate speculation, they also seek to extract profit through the realization of merchandise. The Thomé family lives up to the concept of the land-capital alliance (MARTINS, 1994).

3.4.2 The Solomon family: land, modernity and land income

The Salomão family has a similar history to the Thomé family, as they are also part of the group of Syrian-Lebanese immigrants who arrived in Três Lagoas to work in local commerce and in cattle breeding. According to local media reports, the Salomão family arrived in the municipality in mid-1910, through Jamil Jorge Salomão. It was in Três Lagoas that Jamil started his family, marrying and having two children, whose activities are present in the commerce sector, livestock farming and the setting

---

[37] According to the National Institute for Colonization and Agrarian Reform (INCRA), the de-characterization of rural property as urban property occurs when "a rural property loses its agricultural purpose and becomes part of an urban perimeter". (INCRA, 2022).

up of new allotments and/or gated communities. The current deputy mayor of the municipality, elected in 2016, Paulo Salomão Nery, is his grandson.

The Salomão family stands out as a landowning family because of two fundamental characteristics: the first is their investment in modern developments in the city, such as Shopping Três Lagoas, according to registration 63.078, annex 8, allotments and gated communities; and the second is the fact that the family owns land in other cities in Mato Grosso do Sul. As such, they are well known in the city due to their regional influence.

Traditionally, the Salomão family has maintained an economic focus on cattle breeding and land ownership in the municipality of Três Lagoas, such as Fazenda Vovô Jamil Jorge and Fazenda Santa Helena. Despite the concentration of rural areas, in recent years, especially since 2012, the family has focused its activities on the urban real estate sector, producing allotments and condominiums in partnership with real estate developers.

The Santa Helena farm, the family's main property, is close to the northern part of the city, an area that is experiencing significant growth due to the installation of the Três Lagoas Shopping Center, the airport, the municipal spa and gated communities.

Fazenda Santa Helena, as well as Reflorestamento e Pecuária São Thomé Br, were divided into several plots and, according to the registration of area 63.684, annex 1, of the Real Estate Registry Office of the Três Lagoas District, the Salomão family sold an area of plot A, of 48.400 hectares to the group Z-Incorporações Imobiliárias Ltda, from São Paulo, for the price of R$10,500,000, for the construction of the Villa Dumont residential gated allotment. Dividing the total value by the number of hectares gave a value of R$ 216,942.1487 per hectare.

The family's businesses are mainly in the municipality of Três Lagos, as shown in Table 13, ranging from gas stations to the import and export of wood. Activities that have grown since the territorialization of eucalyptus in the municipality, especially in the real estate business, through the launch of the Santa Helena Residential Allotment, near the Shopping Mall, whose plots are being sold for an initial value of $92,689.33[38]

.

Table 13: Salomão Family enterprises

Company name	Municipalities	CNPJ	Home	Share Capital (R$)
Auto Posto Px Ltda	Três Lagoas-MS	15.380.587/0001-89	07/05/2015	R$ 234.000
J. H. Salomao - Industria, Comercio, Importação e Exportação de Madeiras Ltda	Três Lagoas-MS	07.409.496/0001-09	02/06/2005	00
Allotment Santa Helena Empreendimento Imobiliário Spe Ltda	Três Lagoas-MS	31.795.759/0001-31	18/10/2018	R$ 20.000.000
J. H. Salomão - Industria, Comercio, Importação e Exportação de Madeiras Ltda	Nova Bandeirantes -MT	07.409.496/0001-09	02/06/2005	00
J. H. Salomão - Industria, Comercio, Importação e Exportação de Madeiras Ltda	Três Lagoas-MS	07.409.496/0002-81	26/02/2008	00
Jefferson Jorge Salomão and Others	Três Lagoas-MS	(12.300.084/0001-68)	27/07/2010	00
Jefferson Jorge Salomão and another	Guaraçai-SP	08.159.370/0001-88	13/07/2006	00
Shopping F3 Empreendimentos e Participações Ltda.	Três Lagoas-MS	14.183.259/0001-20	06/07/2011	00
F3 Administradora e Participações Ltda	Três Lagoas-MS	24.691.294/0001-03	29/04/2016	R$ 605.000
Auto Posto Nações Ltda	Três Lagoas-MS	18.906.482/0001-09	11/09/2013	R$ 100.000
Total: R$ 20,939,000				

Source: Brazilian Federal Revenue Service

Another significant undertaking by the Salomão family in the municipality was the construction of Shopping Três Lagoas, also on a 12-hectare portion of Fazenda Santa Helena. The construction area was valued by the public authorities at R$840,000.00 when calculating the payment of the Real Estate Transfer Tax (ITBI) and was carried out by the company Vértico Três Lagoas Empreendimentos Imobiliários de Três Lagoas Ltda. According to information from the local Três

Lagoas media, via the Diário Digital portal[39] , the Shopping F3 Empreendimentos e Participações Ltda project was budgeted at around R$100,000,000.

Although several of the Salomão family's companies don't have share capital declarations, those that do total R$20,939,000 and, if added to the estimated amount invested in the construction of the shopping center, the total jumps to R$120,939,000.

Although Três Lagoas is the municipality where the family lives and sets up its enterprises, the Salomão family owns farms in other municipalities in the state of Mato Grosso do Sul, according to the data in Table 14. The family's remaining areas in Três Lagoas total 588.5935 hectares.

Table 14: Rural properties owned by the Salomão family in Mato Grosso do Sul

Property	Area in hectares	Municipality
Vovo Jamil Jorge Farm	815,1051	Three Lagoons
Fazenda Santa Helena - Gleba A / Fazenda Santa Helena - Gleba A	171,1452	Three Lagoons
Pedra Farm	0,5855	Three Lagoons
Pedra Farm	0,5855	Three Lagoons
Pedra Farm	0,5856	Three Lagoons
Pedra Farm	0,5856	Three Lagoons
Santa Anália Farm / Dismembered Area	88,8657	Paraíso das Águas - MS
Santa Anália Farm / Remaining Area	489,2888	Paraíso das Águas - MS
São Felipe Farm / Dismembered Area	128,5205	Água Clara - MS
Pedra Farm / Dismembered Area	30,5723	Água Clara - MS
Pedra Farm / Dismembered Area - B	411,0946	Água Clara - MS
Pedra Farm / Remaining Area	367,6933	Água Clara - MS
Pedra Farm / Remaining Area	1.656,6055	Água Clara - MS
Pedra Farm	2.067,6980	Água Clara - MS
Santa Anália Farm	578,1550	Chapadão do Sul - MS
São Felício Farm / Gleba A	927,1043	Chapadão do Sul - MS
São Felício Farm / Gleba B	556,9757	Chapadão do Sul - MS
Lagoa Farm - Gleba 02 / Registration 7912	166,4833	Chapadão do Sul - MS
São Felipe Farm	4.433,7962	Porto Murtinho - MS
São Felipe Farm / Dismembered Area	128,5205	Porto Murtinho - MS
São Felipe Farm	4.432,7906	Porto Murtinho - MS
São Felipe Farm	994,8201	Porto Murtinho - MS
São Felipe Farm / Remaining Area	4.305,24	Porto Murtinho - MS
Total area: 20,984.8566		

Source: INCRA Land Management System

---

[39] Available at <https://www.diariodigital.com.br/geral/inaugurado-o-primeiro-shopping-de-tres-lagoas/#:~:text=Or%C3%A7ado%20em%20R%24%20100%20milh%C3%B5es,do%20Sul%20e%20S%C3%A3o%20Paulo.> Accessed on January 27, 2022.

The family owns a small amount of land in the municipality of Três Lagoas, but in the state of Mato Grosso do Sul they own a total of twenty-three rural properties located in the following municipalities: Três Lagoas, Paraíso das Águas, Água Clara, Chapadão do Sul and Porto Murtinho, which together result in 20,984.85 hectares.

In addition to these areas, the family also owns nine other properties in the state of São Paulo, in towns close to Três Lagoas, such as Castilho, Andradina and Guaraçai. The areas of these properties together total 1,900.5567 hectares, as shown in Table 15.

Table 15: Rural properties owned by the Salomão family in the state of São Paulo

Property	Area in hectares	Municipality
São Lourenço Farm / São Lourenço Farm - Dismembered Area	749,8545	Guaraçai
São Lourenço Farm / São Lourenço Farm - Remaining Area	418,2477	Guaraçai
Fazenda São Lourenço / Fazenda São Lourenço - Gleba "F" - Dismembered Area	186,0073	Guaraçai
Fazenda São Lourenço / Fazenda São Lourenço - Gleba "F" - Remaining Area	185,1337	Guaraçai
Fazenda São Lourenço / Fazenda São Lourenço - Gleba "G" - Dismembered Area	41,7328	Guaraçai
Fazenda São Lourenço / Fazenda São Lourenço - Gleba "G" - Remaining Area	112,2652	Guaraçai
Fazenda São Lourenço - Gleba N° 02 / Fazenda São Lourenço - Gleba N° 02	8,7499	Andradina
Fazenda São Lourenço - Gleba N° 06 / Fazenda São Lourenço - Gleba N° 06	53,2323	Murutinga do Sul
São Lourenço Farm / Fumagalli II Farm	145,3333	Guaraçai, Andradina and Murutinga do Sul
Total area: 1,900.5567		

Source: INCRA Land Management System (SIGEF), 2021.

The Salomão family, despite having large urban enterprises in Tres-Lagoense, is representative of rentier capitalism because they have a monopoly on land, both in Mato Grosso do Sul and in the state of São Paulo. The properties in the two states add up to 22,885.41 hectares.

The Salomão family's ownership of several land properties and their luxurious investments in urban areas make them part of the elite that make up the land-capital unit, since they earn their income from land ownership and commercialization, and also make a profit through commerce. The family's economic activities show different ways of earning land income.

The sale of areas dismembered from Fazenda Santa Helena to create the Villa Dumont condominium classifies the realization of absolute land rent, through the sale of the property. According to Marx (2017), when the owner sells his land, he calculates from this the percentage price of income that the property would generate, which resulted in the significant price of R$10,500,000.

The strategy of building the Santa Helena allotment near the Três Lagoas Shopping Center is a way of earning differential income I, since the privileged location of the allotment will be incorporated into the price of the plots for sale. Furthermore, according to Rodrigues (1988), the strategy of incorporating hectares into the urban perimeter and parceling them out in the form of allotments, per m², increases the price of land, since urban land is more profitable.

The Salomão family also participates in activities centered on capitalist agriculture, since, according to data from the Federal Revenue Service, the company Jefferson Jorge Salomão e Outros, located in Guaraçai-SP, operates in the rural activity sector, planting sugar cane. The sugar cane produced on the properties is used to supply raw materials to the Raizen agro-industry, which has one of its headquarters in Andradina, the Gasa Unit.

In this way, the Salomão family is part of the class of landowners who have now diversified their businesses into the service sector, capitalist agriculture and/or those directly related to the dynamics of earning income from the land. They also play a direct role in local politics, through deputy mayor Paulo Salomão Nery, who wants to run for state and federal office, according to the news portal Hojemais Três Lagoas .[40]

These are the characteristics that highlight the family's participation in the land-capital pact. What's more, it encourages us to think about the current complexity of the land-capital pact, since it doesn't just mean the fusion of class interests (landowner and capitalist), especially from the perspective put forward by Martins (1994) in which the capitalist becomes the landowner. In Três Lagoas, the path is also reversed: landowners become urban capitalists with the diversification of the real estate "portfolio" provided by the land monopoly.

3.4.3 The Prata Tibery family: traditionalism, land income and the state

---

[40] Available at <https://www.hojemais.com.br/tres-lagoas/noticia/politica/temos-que-estar-preparados-seja-para-que-cargo-for-diz-paulo-salomao-sobre-disputar-as-eleicoes-deste-ano>   Accessed   on February 05, 2022.

The Prata Tibery family are of Italian descent and began their journey in Brazil in the state of Minas Gerais, in the Triângulo Mineiro region, near Uberaba. The family dedicated themselves to cattle breeding and made a name for themselves in the sector by importing the first zebu from India.

In 1936, part of the Tibery family, through Orestes Prata Tibery, migrated to Três Lagoas, maintaining the essence of their economic activities in cattle breeding. The family gained notorious municipal recognition through Orestes Prata Tibery Junior, who became one of the country's leading cattle breeders, with five children who are continuing the family's enterprises.

The Prata Tibery family has established itself in the municipality of Três Lagoas as one of the elite landowners. Today, however, the family's economic activities are very diversified, mainly concentrated in urban activities, especially in the hotel business and in the construction of new allotments, as shown in Table 16.

Table 16: Tibery Prata Family Enterprises

Company name	Municipality	CNPJ	Date	Share Capital
Hotel OT Ltda	Three Lagoons	02.331.794/0001-00	19/01/1998	R$ 869.250
Central Ms Representações Ltda	Three Lagoons	09.075.084/0001-05	17/09/2007	R$ 5.000
Centro De Eventos Leiloado Ltda	Three Lagoons	04.579.175/0001-56	26/07/2001	R$ 114.000
Auto Posto Jc, Hotel E Restaurante Ltda	Three Lagoons	21.109.967/0001-86	25/09/2014	R$ 235.000
Residencial Império Empreendimento Imobiliário Spe Ltda	Paranaíba	22.851.614/0001-10	15/07/2015	R$ 100.000
Jp-Construção Civil, Engenharia E Projetos De Arquitetura Ltda	Three Lagoons	35.721.280/0001-01	06/12/2019	R$ 10.000
Taj Hotel Ltda	Three Lagoons	08.188.452/0001-50	13/07/2006	R$ 1.456.308
Total: 2,789,558				

Source: Receita Federal, 2021.

Orestes Prata Tibery Junior's family has seven CNPJs, operating in various sectors, such as services, real estate and hotels. The sum of the family's declared share capital is R$2,789,558.

According to fieldwork interviews, the Prata Tibery family sold part of their land to set up residential allotments and gated communities:

> Condominiums on their land include Quarta Lagoa, terras do Jupiá, Bosques de Araras, Set Sul subdivision and Condomínio Portal de Águas; then there are the popular subdivisions that are OT, and the Orestinho and Novo Oeste housing developments, which were part of the Minha Casa Minha Vida program.

The allotments mentioned were built on areas of Tibery's farms, such as Fazenda Vovó Ruthy, registration 58.575, annex 7, where the Orestinho and Novo Oeste housing complexes are located, and the OT allotment, whose land was bought by the town hall, according to registration 62.714, annex 2. The other allotments come from the area of Fazenda Santa Luzia, which was divided into plots and incorporated into the urban perimeter, according to registration 46.928, annex 3, which gave rise, for example, to the Terras de Jupiá allotment, produced by the real estate developer Vectra Construtora Ltda, based in Londrina-PR.

Despite the growth of urban enterprises, especially those related to the practice of earning income from urban land, the family also has economic activities anchored in the concentration of rural properties, as shown in Table 17.

Table 17: Rural properties of the Prata Tibery family

Property name	Area (in hectares)	Municipality
Two Brothers Farm	28,8365	Three Lagoons
Ot Ranch I	8,00	Three Lagoons
Nelorandia Resort	242,00	Three Lagoons
Estancia São Jose	49,76	Three Lagoons
Etn Riacho Doce	6,6	Three Lagoons
Sitio N S Aparecida	10,7	Three Lagoons
Amizade Farm	266,3	Three Lagoons
Vera Cruz Farm	93,1696	Three Lagoons
São Joao Farm Gleba 1 and 2	598,3941	Three Lagoons
Vo Orestes Farm	452,8983	Three Lagoons
Vo Ruth Farm	408,1493	Three Lagoons
Lagoinha Resort	224,2619	Three Lagoons
Faz Santa Helena	17,735	Three Lagoons
Ot Ranch	7,8222	Three Lagoons
Santa Luzia Farm / Santa Luzia Farm - Parcel 01	7,0472	Three Lagoons
Santa Luzia Farm / Santa Luzia Farm - Parcel 02	2,7236	Three Lagoons
Santa Luzia Farm / Parcel 3	465,4272	Three Lagoons
Santa Luzia Farm	467,9402	Three Lagoons
Santa Luzia Farm	477,5691	Three Lagoons
Estância 5 Irmãos / Part 1	139,1147	Three Lagoons
Estancia 5 Irmãos	178,2185	Three Lagoons
Estancia 5 Irmãos	39,1038	Three Lagoons
Vista Alegre Da Cachoeira Farm	136,7124	Ribas do Rio Pardo - MS

Total area: 5,194.5015

Source: INCRA Land Management System (SIGEF), 2021.

According to SIGEF data, the family owns twenty-four properties, twenty-two of which are in the municipality of Três Lagoas and total 4,636.99 hectares. The other two properties are located in the municipality of Ribas do Rio Pardo and total 557.50 hectares. The twenty-four areas add up to 5,194.50 hectares.

In the case of the Tibery family, bearing the surname goes beyond economic relations, becoming part of the identity of being part of a distinct group of powerful people with a territorial reference. The children continue to carry on the activities historically carried out by the family, such as raising Zebu cattle.

The various ventures on Três Lagoas soil consolidated the Tibery family's notorious local recognition, especially in the media's statements about Orestes Prata Tibery Junior. According to Perfil News reports from Três Lagoas[41] , he was considered an "enlightened person" and "a visionary man".

Orestes Prata Tibery Junior did not dedicate himself directly to political life, but he did maintain a relationship of dialog with important names in Brazilian politics, as he emphasized in his interview with the RCN 67 *website*[42] about his relationship with former president Lula (PT):

> President Lula was a fantastic partner in helping us import the embryos from India, interfering directly on two occasions. Thanks to him, we now have products from these embryos born in Brazil (RCN News Portal 67).

The account of his relationship with former president Lula (PT) reinforces the understanding of the realization of private class interests in Brazilian politics. The landowning class, even when they don't participate directly in politics, can be part of economic decision-making and have their interests privileged because they are seen as representing the class. In the debate on the relationship between economics and politics, Martins (1994) states:

> In fact, the indications suggest that political clientelism has always been and is, above all, preferably a relationship of exchanging

---

[41] Available at <https://www.perfilnews.com.br/morte-de-orestinho-completa-cinco-anos-mas-seu-legado-permanece-vivo/> Accessed on: January 27, 2022.

[42] Available at <https://www.rcn67.com.br/jpnews/tres-lagoas/orestes-prata-tibery-jr-um-poeta-zebuzeiro/34802/> Accessed on: January 28, 2022.

political favors for economic benefits, no matter on what scale. It is therefore essentially a relationship between the powerful and the rich, and not primarily a relationship between the rich and the poor (p.29).

The fact is that the powerful, whether they are direct participants in the state or not, have their interests served in order to achieve their class reproduction. An example of this is the compensation that the Prata Tibery family received from CESP during the construction of the hydroelectric plant.

During the construction of the hydroelectric plant, some areas on the banks of the Paraná River were expropriated by the state through purchase. According to the interviewee, the CES compensation for this case was voluminous at the time - more than ten million - and was one of the reasons that boosted the family's wealth:

> He was able to leverage the compensation for their land, which had been expropriated by the hydroelectric plant. Through political influence, he was able to negotiate compensation for the hydroelectric plant. It was one of the biggest. He succeeded because his land was on the banks of the Paraná River. He received compensation from CESP, which was one of the largest, at the time more than ten million. A huge amount of money he received in compensation from CESP. He got a lot of money out of it (Interviewee A - 03/02/2022).

Although the area has been indemnified by the State, according to the administrative improbity process, No. 0802782492015812002[43] , of 2015, of the Public Prosecutor's Office of the State of Mato Grosso do Sul, whose defendants are Orestes Prata Tibery Junior and five CESP employees, respectively: Guilherme Auguto Ciner de Toleto - former President; José Aparecido de Lira - Manager of the Legal Department of the interior; Carlos Educado Epaminondas França and Reinaldo José Rodriguez de Campos - Administrative Directors, continued to be used by the former owner, in this case, Orestes Prata Tibery Junior, according to:

> Finally, Orestes Prata Tibery Júnior, for his part, placed his private interest above the public interest and continued to use the area, for which he had already received the significant sum of more than twelve million reais, free of charge, even resisting CESP's attempts - albeit late - to establish a cost for the transfer (folios 13/45 and 83/84). He was the direct beneficiary of the damaging act (Case no. 0802782492015812002, p.05).

---

[43] Administrative improbity case no. 0802782492015812002, of 2015, from the Mato Grosso do Sul State Public Prosecutor's Office.

According to the lawsuit, although the area was expropriated by CESP, the farmer continued to use it to raise cattle, with the approval of CESP officials. According to the case file, the officials justified that the farmer had managed to document the use of the area through the CESP concession, however, no documents to this effect were found in the investigations.

According to the arguments of public prosecutor Antônio Carlos Garcia de Oliveira, the expropriation was a facade so that the owner could receive the compensation, which in total amounted to R$12,069,614.30. However, in practice, the area remained in the possession and use of the owner, according to:

> Remember that in the case in question, as has been said, there is no doubt that Orestes Prata Tibery was awarded the value of an expropriation that earned him millions (more than R$12 million) and, amazingly, he didn't need to dispose of the area. In practice, the expropriation didn't take place, as he used the expropriated area as normal. The expropriation was just a façade to allow him to receive the millionaire sum (n° 0802782492015812002, p.8).

In addition to this process, which involves benefiting one's own interests to the detriment of public interests, Orestes Prata Tibery Junior's sister, Wania Ruth Prata Tibery, is also responding to a process, still under analysis, of overlapping private and public areas. The area corresponds to 12.5813 hectares and is part of the so-called Fazenda Santa Luzia - Gleba B.

The complaint was made by Stella Marques de Souza Zopff, in Case No. 08039173320148120021[44] , of 2014, of the Public Prosecutor's Office of the state of Mato Grosso do Sul, requesting the interdiction of the area, since the Prata Tibery family has exceeded the limits of the fence into their property, according to:

> On July 16, 2014, a fence was erected in front of the area owned by the plaintiff, but without actually fencing off any area; afterwards, on July 17, 2014, the plaintiff received a call informing her that a tractor was on her property tearing down her boundary fences. The plaintiff then went to the site and found the tractor on the side street, and when she approached the workers who were there with the tractor, they said that they were doing the work on the orders of Mrs. Wania, now a partner and owner of the defendant company. (Case no. 08039173320148120021, p. 3).

Following the complaint, the area was investigated and the state Public Prosecutor's Office found that in addition to the incorporation of a private area, the

---

[44] Invasion of land case no. 08039173320148120021, of 2014, by the Mato Grosso do Sul State Public Prosecutor's Office.

Prata Tibery included public areas in its perimeter and registered them on the property's license plate:

> There are also other irregularities in the rectification, which render it null and void, such as the area rectified being part of a public road, among others already mentioned. This administrative rectification has entered a public area, where it encroaches on part of the following streets: Rua Turmalina (facing block 247, of the Jardim Alvorada allotment); part of the old Rua Andradina now called Rua João Gonçalves de Oliveira (side of blocks 240 and 352, of the Jardim Alvorada allotment); part of Rua Bom Jesus da Lapa (side with blocks 240 and 247, of the Jardim Alvorada allotment); part of the former Rua Aquidauna, today called Rua Bernardino Rodrigues Montalvão (side of block 247, of the allotment called Jardim Alvorada); in other words, an impossible act, as it is an affected area; all of this observing the map of the allotment called Jardim Alvorada. (nº 08039173320148120021, p. 292).

The process is still ongoing, however, and as well as the plaintiff denouncing the overlapping of the area, she is also requesting the nullity of the registration, since other problems have been found in relation to this area, such as the incorporation of public land.

Martins (1994) argues that Brazilian oligarchism is not based solely on clientelist relations, of favors between antagonistic and complementary classes.

> My conception is that Brazilian oligarchism is based on something broader than this relationship - it is based on the institution of political representation and on the state. Not only the poor, but all those who in some way depend on the state, are induced into a relationship of exchanging favors with politicians (MARTINS, 1994, p. 29).

According to the author, the state and/or the representatives who make up the state machine serve the interests of the dominant classes, whether industrial capitalists or landowners. Historically, these relationships are the essence of national politics, the center of which is the patrimonialist state, based on clientelist relationships. The case of Três Lagoas is further evidence of what is most solid in the heritage of the constitution of the national state, the intertwining of public and private interests as a tool for the enrichment of the powerful.

3.5 City under siege: traditional families, land rents and rising land prices

The urban perimeter of Três Lagoas was institutionalized by means of registration 462, annex 4, on June 22, 1922, which transferred 3,659 from the State of Mato Grosso to the municipality, for the purpose of the settlement of Três Lagoas, via State Presidency Decree no. 311 of April 9, 1912, according to:

> The Registry Officer. (as) Felipe Nery Monteiro. **Name, Domicile, Profession, State and Residence of the Purchaser: THE MUNICIPALITY OF TRÊS LAGOAS**, of the State of Mato Grosso, with headquarters in the city of Três Lagoas. **Name, Domicile, Profession, State and Residence of the Transferor: THE STATE OF MATO GROSSO**, with its capital in the city of Cuiabá. **Title of transfer**: Concession. **Form of Title, Date and Registrar:** Definitive title issued on March 13, 1992 by the state's Land, Mining and Colonization Department, signed by the Department's Director, surveyor Otavio de Vasconcelos Neves, and signed by the President of the State, Colonel Pedro Celestino Corrêa, and the Secretary of Agriculture, Dr. Carlos Borralho. **Contract value:** None. **Conditions of the Contract:** The lot granted was to serve as a patrimony for the settlement of Três Lagoas, which was being formed at the time of the concession (Matrícula 462. Livro 3-A, fls. 112).

The perimeter created corresponded to the following location:

> Bordering to the north on the property of the Feira do Gado de Três Lagoas company and on the Varginha farm belonging to the estate of Delfino Pereira dos Santos; to the west on the Córrego do Pinto farm, belonging to several people; to the south on the bed of the Noroeste Brasil railroad and to the east on the aforementioned property of the Feira do Gado de Três Lagoas company, the perimeter being demarcated with (50) fifty markers arranged in the manner stated in the title.

We have tried to delineate the area in question in a figure so that it can be better understood, as shown in map 6. The area delimited as the initial perimeter of the municipality of Três Lagoas was used for settlement purposes, through the subdivision and sale of plots.

The immediate vicinity of the perimeter is the property of Companhia Feira do Gado de Três Lagoas, dismembered from Fazenda Santa Helena, which was previously an area owned by the state of Mato Grosso of around 4,050 hectares, according to registration number 2,385, annex 5:

> **Order No.: 2.385. Date:** November 28, 1936. **Circumscription:** Municipality of Três Lagoas. **Title or Street and Property Number:** "Santa Helena". **Characteristics and Confrontations:** rural property called "Santa Helena", with an area of four thousand and fifty hectares (4,050) (Matrícula 2.385, Book 3-F, fols. 33).

Part of this area, which is not included in the registrations, was sold by the state of Mato Grosso to Companhia Feira do Gado de Três Lagoas for the significantly low price of Rs. 200,000 (two hundred contos de reis)[45] . According to the license plate:

> **Name, Address, Profession, State and Residence of Purchaser:** **COMPANHIA FEIRA DO GADO DE TRÊS LAGOAS**, in liquidation, with registered office in the city of São Paulo. **Name, Domicile, Profession, State and Residence of the Transferor: THE STATE OF MATO GROSSO. Title of transfer:** Definitive title. **Form of Title, Date and Servicer:** Definitive title issued by the Secretary of Lands and Public Works on November 17, 1936, and signed by Governor Dr. Mário Correa da Costa and M. C. Oliveira Mello. **Contract value:** Rs. 200:000 (two hundred contos de reis). **Conditions of the Contract:** The land laws (Matrícula 2.385, Book 3-F, folio 33).

The initial perimeter of Fazenda Santa Helena was not located using the information from the Real Estate Registry Office of the Três Lagoas District, however, the area corresponding to the sale made by the state of Mato Grosso to Companhia Feira do Gado de Três Lagoas corresponds to the perimeter also demarcated on map 6.

Companhia Feira do Gado de Três Lagoas, which was responsible for buying part of Fazenda Santa Helena, had its headquarters in the city of São Paulo, which at the time was the center of trade relations established in agriculture. It is not known, however, who owned this company, whether they used it to raise cattle or as a store of value.

The probable hypothesis is that Fazenda Santa Helena, which originated in the state of Mato Grosso, was divided up and sold in Glebas, according to the remaining area of Fazenda Santa Helena - Gleba A, which is listed as a possession of the Salomão family, also located on map 6. In addition to Gleba A, other possessions also originated from this area, such as the perimeter of Fazenda Reflorestamento e Pecuária São Thomé Br, dismembered from Companhia Feira do Gado de Três Lagoas, according to map 6.

Map 5 - Três Lagoas (MS): Representation of the initial perimeter of the city and the rural properties Santa Helena, Vó Ruth, Reflorestamento e Pecuária São Thomé Br and Companhia Feira do Gado de Três Lagoas

---

[45] The corresponding price in reais would today be around R$4,200.00, according to information from Cédulas BR, available at <https://www.cedulasbr.com.br/index.php/2012-11-07-17-53-43/2012-11-07-18-05-59/2012-11-08-16-28-25> Accessed on February 5, 2022.

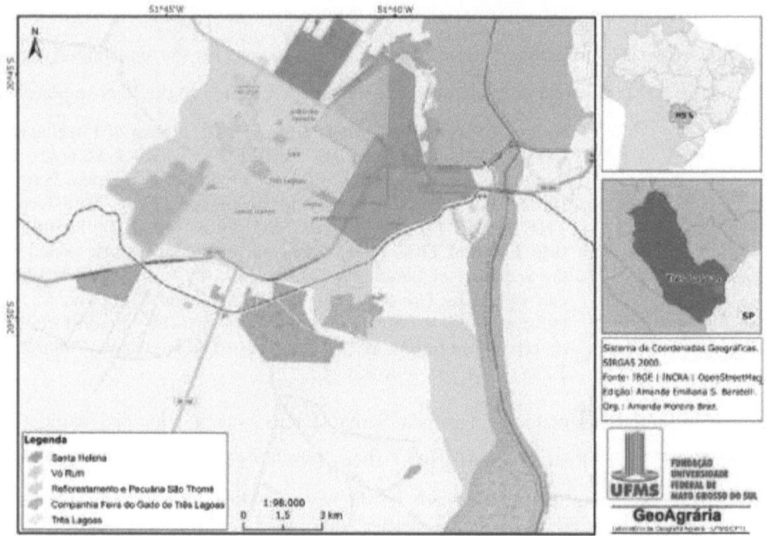

The initial perimeter for the creation of the municipality of Três Lagoas for settlement purposes, via the sale of plots, gave rise to a land monopoly, as evidenced by the ownership of Fazenda Vó Ruth, owned by the Tibery family.

The areas selected as possessions of the Thomé, Salomão and Prata Tibery families surround the city of Três Lagoas in the north and west regions. These areas have recently been reducing their rural dimensions and being incorporated into the urban perimeter to form new allotments.

Map 7 shows the current perimeter of these areas. Part of these possessions have been incorporated into the city's urban perimeter.

Map 6 - Três Lagoas (MS): Current representation of the rural properties Fazenda Santa Helena, Fazenda Vó Ruth and Fazenda Reflorestamento e Pecuária São Thomé - 2022

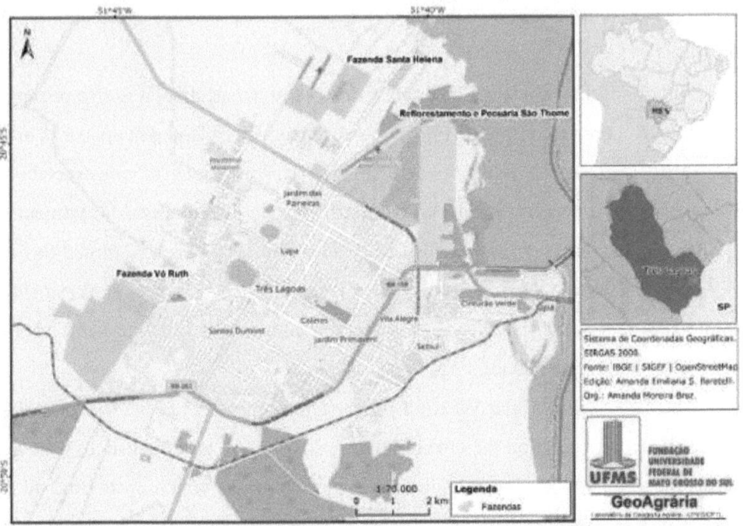

Today, the dimensions incorporated into the urban area form residential subdivisions, such as the Orestinho and Novo Oeste housing estates, resulting from the de-characterization of the Vó Ruth farm, owned by the Prata Tibery family.

Fazenda Santa Helena - Gleba A, corresponds to the area where Shopping Três Lagoas, Villa Dumont gated community and the Santa Helena subdivision are located today.

The Fazenda Reflorestamento e Pecuária São Thomé was divided in half by the Samir Thomé Ring Road, on one side of which were the Paranapungá and Acácias neighborhoods, the Três Lagoas Regional Hospital and the area with a rental advertisement - indicated in the previous topic 4.4.2.

The Thomé, Salomão and Prata Tibery families mirror the rentier capitalism materialized in the appropriation of differential income I and II, the result of the process of reserving the value of land (monopoly), as they waited for the ideal moment to be incorporated into the urban perimeter. These processes have taken place over the last decade and are related to the dynamic increase in land prices resulting from the territorialization of eucalyptus.

In this way, the dynamics of land price speculation, especially when it occurs through the subdivision of plots in urban areas, aims to realize absolute rent, since these properties will be sold with prices increased by the percentage of differential rent I and II.

Although Marx analyzed land rent from a rural perspective, it is also present in urban areas. According to Lojkine (1981), three categories of land rent appear in urban areas: differential rent I, which occurs in different situations of constructability; differential rent II, which is produced by the difference in capitalists' investments in shopping centers and real estate; and, finally, absolute rent, which is earned through the existence of private land ownership. This is because, according to Harvey (1980), absolute land rent is the rent charged for the simple fact of occupying land, whether for existence and/or production.

The sale of part of the Vó Ruth property took place between 2010 and 2012, when construction began on the Orestinho and Novo Oeste housing estates. The area sold by the family was divided into three plots, two of which were sold to the government and one of which was concentrated among the popular condominiums, acting as a store of value for real estate speculation.

According to Rodrigues (1988), real estate speculation occurs in two ways, one of which is speculation in plots, related to subdivision and sale. The strategy of this form of speculation is the initial sale of plots in areas with little urban infrastructure and the concentration of "empty" plots closer to the service sectors. As the subdivision expands and gains population density, the services come closer and, consequently, the plots closest to them tend to increase in value.

The other form of speculation is called "gleba" speculation. A gleba is a large piece of land that divides two subdivisions. While the neighboring subdivisions expand and attract services, the gleba concentrated between them is "waiting" for its value to rise so that it can be parceled out. In this sense, in addition to the private services added to the region, the state also acts with improvements and basic infrastructure, which is called the social production of the urban.

The area that was used for real estate speculation between the years when the housing complexes were consolidated and, above all, when the infrastructure was installed, is now being sold off by the Impper real estate development group. According to the information in the Property Registration 62.714, annex 2, the area for the construction of the OT Allotment (Orestes Prata Tibery) was sold for the price of R$ 7,000,000.00.

Image 5: Advertisement for the sale of plots in the Orestes Prata Tibery subdivision

Source: Fieldwork, 2022. **Photo:** Baratelli, A. E. S., January 15, 2022.

In addition to the presence of the new allotment on the Prata Tibery family's land, there is also the construction of the Império residential development, which will function as a gated community, as shown in images 6 and 7.

Image 6: Advertisement for the Residencial Império gated subdivision

Source: Fieldwork, 2022. **Photo:** Baratelli, A. E. S., January 15, 2022.
Image 7: Império Residential Condominium

Source: Fieldwork, 2022. **Photo:** Baratelli, A. E. S., January 15, 2022.

On the Impper group's *website*[46] , the advertisements for sale refer to the quality of the infrastructure already in place on the plot. Activities such as earthworks, electrical installation, household connections, paving, sewage system, traffic signs, gutter guides, drainage and landscaping are already 100% complete, as shown in image 8.

Image 8: Construction schedule for the OT subdivision

Source: Impper Group Official *Website*

<hr>

46 Available at <https://grupoimpper.com.br/empreendimentos/residencial-ot/> Accessed on February 05, 2022.

In addition to these infrastructures that have already been consolidated in the subdivision, the developer's advertising also includes the presence of assets of public interest, such as: a community square, a future avenue leading to the center, residential and commercial plots, a nursery school and a future municipal school, a future UBS and a future CRAS, complete infrastructure for water, electricity, asphalt and sewage, etc., as shown in image 9:

Image 9: Advertisement using the publics incorporated into the OT Allotment

Source: Impper Group Official *Website*

The fact is that all of these infrastructures were built by the government during the period in which the housing complexes were set up. However, they are being used as a way of earning differential income II, by incorporating the surrounding infrastructure to increase the price of the plots.

The same dynamics of speculation were carried out by the Salomão family, who started building the Três Lagoas shopping center in 2012 and are currently producing the Santa Helena subdivision, which has similar characteristics.

The subdivision in question is located next to Shopping Três Lagoas, as shown in map 8, close to the airport and the municipal spa. There are also luxury gated communities on this perimeter, such as Villa Dumont - also on land dismembered from Fazenda Santa Helena. These territorial images are ideal for selling the idea that "the subdivision will offer all the security you need to live in a neighborhood with gated community characteristics".

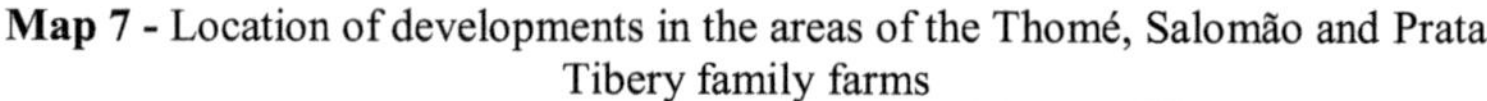

Map 7 - Location of developments in the areas of the Thomé, Salomão and Prata Tibery family farms

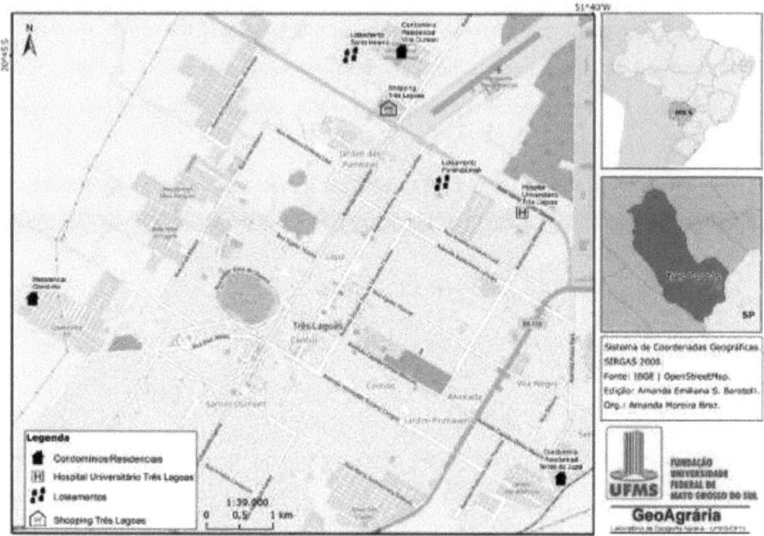

The Santa Helena subdivision, even though it has not yet been plotted, already has basic urban infrastructure in place. According to information on the official *website*[47] , the stages of opening up the streets, building the river galleries, installing the water and sewage systems and the gutter guides are already 100% complete, as shown in image 10. The streets and avenues leading to the subdivision have been completely paved.

---

[47] Ramos Junior Engineering. Residencial Santa Helena. Available at <http://engenhariaramosjunior.com.br/empreendimentos/santa-helena/> Accessed on: October 30, 2021.

Image 10: Allotment project for Santa Helena

Source: Official *website* of the Santa Helena allotment

According to Rodrigues (1988), the installation of urban infrastructure, represented by basic facilities, systematically increases the price of land. In this way, the rapid installation of basic urban facilities for the launch of new subdivisions shows the presence of the government as an agent/partner of real estate speculation.

The installation of the shopping center on the way to the airport and the municipal spa has the public authorities as the ideal partner to install the necessary infrastructure for the municipal flow, such as paving, a sewage system, electrical infrastructure, etc. Thus, in the process of privately marketing the plots in the new subdivision, the improvements made by the state will be incorporated into the land in the form of differential rent II.

Like the Prata Tibery and Salomão families, the Thomé family has also exercised and continues to exercise rentier practices in the municipality of Três Lagoas, since part of their Fazenda Reflorestamento e Pecuária São Thomé Br was incorporated into the urban perimeter to create the neighborhood. And now, with areas donated to the Três Lagoas Regional Hospital, whose structure is almost ready, it will

135

also have the possibility of incorporating differential income II in the sale and/or lease of areas.

When we asked the interviewees whether the new capitalist dynamics introduced in the municipality of Três Lagoas, due to the expansion of eucalyptus monocultures, had changed local power relations and/or hindered the traditional elite's ventures, we obtained the following answers that corroborate our analysis:

> In reality, what has happened to the city is that, at the beginning, it was basically built on farming and cattle-raising, but with the recent industrial development of the city, they saw the opportunity to make more money. They haven't lost out, quite the opposite, these allotments are made over the long term, producing profit and income over the long term. (Interviewee A - 03/02/2022).
>
> They're splitting if they want to make money. They bought to speculate. They bought by the bushel and sell by the m². (Interviewee B - 24/01/2022) .[48]

In short, these traditionally rentier families saw in the territorialization of eucalyptus the possibility of grabbing a higher percentage of land income, now in the urban area. That's why, in the recent period, they have started to incorporate their rural land into the urban perimeter, diversifying their economic sources and setting up new allotments and urban developments.

3.6 New subdivisions and old subdivisions: the dynamics of contrasts in the "national pulp capital"

The installation of new allotments whose infrastructures have been built by the public authorities[49] are too advanced, highlighting the contrast between class inequality in access to urban improvements in the municipality of Três Lagoas.

The presence of the territorialized cellulose-paper complex in Três Lagoas has significantly altered the economic dynamics of the municipality. The national prominence of these large enterprises, such as Suzano Celulose S/A, Eldorado Brasil and International Paper/IP, attracted national and international attention to the region.

---

[48] Interview conducted on January 24, 2022 with the former MS civil servant.
[49] Available at <https://www.rcn67.com.br/jpnews/tres-lagoas/governo-e-prefeitura-assinam-convenio-de-r-75-milhoes/152346/> Accessed on February 8, 2022.

This is because pulp companies have advanced in eucalyptus monoculture and pulp exports and have stood out in the economy of Mato Grosso do Sul and Brazil. According to Perpetua (2012), in 2009, the former Fibria, now Suzano Celulose S/A, was ranked as the largest pulp exporter in the state of Mato Grosso do Sul. In addition, the volume of wood logs sold in tons is the 3rd most exported product in the state. Furthermore, in terms of national data, the pulp sector in Três Lagoas ranks 56th in the national export *ranking*.

It was these motivations that led Luciano Coutinho, who chaired the BNDES between 2007 and 2016, to classify the pulp sector as the "national champions", i.e. one of the fastest growing sectors in Brazil (PERPETUA, 2012). Speeches like Coutinho's create the ideologization of the pulp sector as independent and responsible for its outstanding international performance. However, it's worth pointing out that the sector has international results and carries out the expanded reproduction of capital due to the voluminous public resources invested in it and, what's more, with co-participation from the state.

Thus, Perpetua (2012) showed that in Fibria alone - now Suzano Celulose S/A - the BNDES owned 30.42% of the shares. Furthermore, the BNDES was willing to subsidize 70.9% of all the investment planned for the pulp sector between 2007 and 2010, or R$11.7 billion. However, the ideology of the sector's growth is based on the discourse of productivity and merit, as if it were independent of state funding.

It was on the basis of the argument that the sector represents the country internationally as a "national champion" that Três Lagoas, due to its prominence in the production of bleached hardwood pulp, was given the title of "national pulp capital". The nomenclature is the result of Senate Bill 178[50] , of 2016, by Senator Simone Tebet (MDB-MS), daughter of former governor Ramez Tebet, whose surname is part of one of the main families involved in politics in the Três Lagoas region. The bill in question was approved in April 2021 and became Law No. 14,142 . [51]

According to the Senator, in the bill's justification, the municipality had impressive data as positive results of the territorialization of pulp enterprises. According to her:

---

[50] Available at <https://www25.senado.leg.br/web/atividade/materias/-/materia/125614>. Accessed on October 25, 2021.
[51]   Available   at   <http://www.planalto.gov.br/ccivil_03/_Ato2019-2022/2021/Lei/L14142.htm>. Accessed on: October 25, 2021.

> The evolution of the data is impressive. During the construction of
> the first plant alone, from 2006 to 2009, more than 20,000 direct and
> indirect jobs were generated; the municipal GDP increased by
> 300%; the state GDP by 13%. (PLS no. 178, 2016, p. 2).

The data presented by the Senator is at odds with the reality experienced by the population of Três Lagoas and also with the surveys carried out in the municipality. Perpetua (2012) points out that, with regard to job creation, the promise made to the population was that around 2,000 to 2,500 jobs would be created directly. While indirect jobs could add up to a maximum of around 10,000 and not 20,000, as the Senator pointed out.

According to City Hall data, between 2005 and 2009, direct jobs amounted to a maximum of 2,300, with Fibria generating 1,300 jobs and Eldorado 1,000. Indirect jobs have not been quantified, but it is known that there has been a 70% increase in employment in the region. However, it can be seen that jobs in outsourced companies show high turnover and signs of seasonal hiring (PERPETUA, 2012). The flexibility of the outsourced companies in hiring/firing in short periods of time represents a problem for the municipality, since the advertisements for vacancies have a national reach and intensify the movement of migration to the municipality and, once fired, the groups of workers find it difficult to maintain the reproduction of life in the city.

What's more, according to the fieldwork, the search for employment by the population of Tres-Lagoense continues to be fraught with obstacles. Even for those with a certain level of education, there are difficulties.

> The cons are the difficulty of registering, the difficulty of living here
> in Três Lagoas and **not being able to get a job**... it's hard to say you
> live here and for outsiders to be hired and you're not. Like, I was an
> administration and HR technician before becoming a manicurist. I
> haven't been able to get anything in my field, I've been posting CVs
> everywhere for four years.
> My husband is a maintenance mechanic, he got a job in Castilho, but
> he couldn't do it here. So those are the cons of Três Lagoas.
> (Interviewee C. My emphasis) .[52]

In addition to employability, the senator's project raises other data that emphasize the economic growth of the population resulting from the installation of the pulp and paper complex. Thus:

> According to the IBGE, between 2009 and 2013, the number of
> salaried workers in Três Lagoas increased by 87.6%. There was also

---

[52] Interview conducted on November 25, 2019 with a member of the Popular Struggle Movement of Três Lagoas.

> a significant impact on workers' incomes: the average monthly
> salary in the same period increased by 14.8%, rising from 2.7
> minimum wages to 3.1 minimum wages. This growth is also
> reflected in the number of companies in the municipality, which
> grew by 27.9%, from 2,597 to 3,322 during this period. (PLS no.
> 178, 2016, p. 2).

The data presented is generalizing and homogenizes the financial condition of the population, not reflecting the reality experienced by the people of Tres-Lago. The Senator's comments do not take into account the excluded/unemployed population. The presence of improved conditions for some selective groups, especially those who are hired to perform administrative and/or planning services, when compared to the general population, reveals the poor distribution of income.

The senator also concludes that the investments made in the municipality are the result of the territorialization of the pulp and paper complex, motivated by the companies' belief in "their people", as she puts it:

> The volume of investments made in the municipality in recent years
> - notably by pulp companies - is eloquent testimony to the fact that
> entrepreneurs believe in the city's economic vocation and in the
> creativity and productivity of its people. (PLS no. 178, 2016, p. 4).

However, the investments mentioned by the Senator do not reach the most impoverished groups of the population, those who live the social ills of a city whose production of wealth is globalized. To understand the size of the social gap between parts of the population of Tres-Lago, we highlight some urban social indicators.

According to Jannuzzi (2009), social indicators have substantive meanings and are used to help understand social reality. Social indicators are divided into two groups, quantitative and qualitative, which allow for different forms of analysis. Quantitative data is generally constructed from public statistics, based on empirical data. Qualitative data, on the other hand, is the result of evaluations by individuals, through public opinion polls or discussion groups. Thus, even when dealing with the same subject, the way in which the data was collected - quantitatively or qualitatively - the results can be different.

In this sense, statements like those made by Senator Simone Tebet in her PL - based on quantitative data from the IBGE - may not actually explain reality. According to data from IBGE Cidades, the GDP *per capita of* the municipality of Três Lagoas is R$96,639.64 and the average income of the population is 2.8 minimum wages. As a result, the city has a satisfactory Municipal Human Development Index (MHDI), which stands at 0.74 - 4th in the state *ranking*.

However, the data presented hides the reality of the excluded people who live in the municipality, because, according to Jannuzzi (2009), they are part of a simple combination of social indicators, such as those used to generate the MHDI, in which education, health and income are analyzed.

There is also research carried out in the municipality of Três Lagoas, through the Sustainable Cities Network, by SEBRAE (Brazilian Micro and Small Business Support Service), published in 2016, whose methodology was based on "semaphore" the city with the colors green (positive criteria), yellow (alert situation) and red (critical situation), in the following services: environmental sustainability and climate change; urban sustainability; fiscal sustainability and governance and competitiveness module.

From the point of view of environmental sustainability and climate change, the most alarming data, considered to be on alert (yellow), was solid waste management which, according to 2015 data, around 93% of waste goes to landfill, including plastics, glass, metals etc., and only 3% is separated for recycling. Noise was also considered to be on alert (yellow), since 15% of those interviewed in the opinion poll complained about the high level of noise in their neighborhoods.

In the area of urban sustainability, the main problem found, in critical condition (red), was sanitation and drainage. According to the data collected, only 47% of households have a sewage system and only 30.02% of wastewater is treated according to national standards. In addition to this issue, mobility and transportation are also in a critical situation (red). The data showed that 85% of the population keeps away from using intra-urban transport, although workers from neighborhoods further away from the center and students use it continuously. The municipality also has a high number of traffic accidents.

According to the RCN 67 news portal[53] , in April 2021, public transportation was suspended by the city government and caused outrage among the population. A resident of the Vila Piloto neighborhood, located in the east of the city, as seen in figure 6, stressed the difficulty of getting around without public transport, due to high fuel prices and excessive heat, making it impossible to use a bicycle.

---

[53] Available at <https://www.rcn67.com.br/jpnews/tres-lagoas/tres-lagoas-segue-sem-o-servico-de-transporte-publico/154854/> Accessed October 26, 2021.

Despite complaints from some of the population, the Director of Transit, Flávio Thomé, said that 35,000 people a month used the transport. He explains that the number is very small and that the fact that the city is flat makes it easier to use electric or human-powered bicycles. As a result, the municipality will continue without urban transport until October 2021.

Other indicators of urban sustainability are on alert (yellow), such as land use and land-use planning, which showed that 45.03% of homes do not meet the country's habitability standards. Many have unfinished construction and/or no walls. The urban inequality data showed that 4.85% of the population is below the poverty line. The data is combined with the presence of social movements fighting for housing, due to the difficulty of paying rent and other expenses.

Still on this topic, education, health and safety data are on alert (yellow). The report showed that 63.3% of teenagers aged 16 to 18 are enrolled in school, while 36.97% have dropped out. In health, it was found that the municipality has 196.3 beds and 161.9 doctors per 100,000 inhabitants, figures that are considered low.

The data presented relates to issues that are part of the daily life of the population of Tres-Lago and are neglected by the public authorities. The presence of poor income distribution and the contradictions are alarming when we analyze the City Hall's revenue, which has grown in recent years. According to data from the Secretariat for the Environment, Economic Development, Production and Family Farming (SEMAGRO), the amount resulting from the Tax on the Circulation of Goods and Services (ICMS), from industry alone, was R$ 163,515,262.79 in 2019. The municipality's own revenues totaled R$167,442,463.00, also in 2019, and the GDP in 2018 totaled R$11,545,054.14.

The high amounts collected by the City Hall, together with the social indicators provided by the IBGE, contribute to the understanding that the municipality of Três Lagoas promotes development and satisfactory social conditions for its population. However, when it comes to analyzing data and information in more depth, as SEBRAE has done, it turns out that the resources collected by the government are not being used to solve the population's problems. It should also be noted that the "large" investments made by pulp companies for "their people" are not reaching those who need them most.

In order to understand whether the problems indicated by the SEBRAE survey still exist, we carried out an investigation using composite data, with interviews and fieldwork, which allowed us to analyze part of the city of Três Lagoas from another perspective, highlighting, above all, the neglect of public authorities in neighborhoods of extreme need.

One of the main characteristics of the city of Três Lagoas is its "sprawl", with neighborhoods on the edges of the city bringing together part of the poor population. Thus, the neighborhoods selected for further analysis are located in the southwest of the municipality and are: Maristela, Chácara Eldorado and Chácara Imperial, as highlighted in red on map 9. Within these neighborhoods are the residential subdivisions: Maristela, Vila Verde; and the housing developments: Violeta I, Violeta II, Jardim Eldorado, Jardim Imperial and Jardim das Violetas.

Map 8 - Três Lagoas (MS): location of neighborhoods

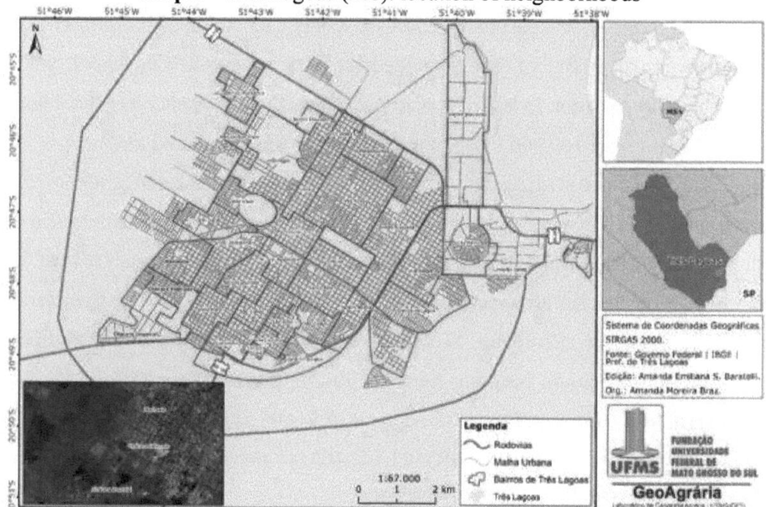

The allotments and housing developments in the selected neighborhoods are very close to each other, so they share the same problems. They appear in the 2016 Territorial Diagnosis - Revision of the Participatory Master Plan (PDP), with various indications about the lack of services in these areas. They were therefore classified as Zones of Social Interest (ZEIS).

> The Special Zones of Social Interest (ZEIS) are areas intended primarily for land regularization, investments in urbanization and the production of housing for the low-income population (PDP, 2016, p. 71).

The Maristela neighborhood appears, in the public opinion survey of the Territorial Diagnosis - Revision of the Participatory Master Plan (PDP), as an area with difficult access to Social Programs, with a lack of paving and a precarious water supply. Violetas I and II was considered by the population to be a precarious area to live in, because it lacks drainage, paving and flooding. The interviewees who live in this area also gave their accounts, as follows:

> [What are the main problems of the residents of this locality?]
> Whether you walk or cycle, I think cycling is the most terrible, because the streets are full of potholes and there are places where there's a lot of dirt, because there's no asphalt here. And when it rains, the dirt comes down onto the only paved street, Manoel Farias Duque. One day, on my way to college, I had to run after my slippers because there's a wide dirt road that divides Vila Verde from Jardim Eldorado, and when it rains... because the bus stop wasn't covered, so when it rained we had to go to the grocery store, because it's covered, it's in front of the school. When it was raining, you had to cross that part of the road that was flooded, so you'd lose your slippers and have to be careful not to fall in, because there were a lot of holes. (Interviewee D) .[54]

The fieldwork allowed us to illustrate the problems experienced by the population, as shown in images 11 and 12, which show the situation of the streets in the neighborhoods in question. In most cases, the streets are difficult to move around in, as they are covered in dirt, potholes, water and gravel.

Image 11: Três Lagoas (MS): side street to Manoel Farias Duque, in the Maristela neighborhood

---

[54] Interview conducted on October 10, 2021 with a resident of the Vila Verde neighborhood.

Source: Fieldwork, 2021. **Photo**: Baratelli, A. E. S., October 25, 2021.

Image 12: Três Lagoas (MS): street in the Jardim das Violetas neighborhood

Source: Fieldwork, 2021. **Photo**: Baratelli, A. E. S., October 25, 2021.

One of the residents' main complaints is about the paving system, since, according to them, there is no planning for the work, so the paved streets don't follow a complete route.

> Asphalt hasn't reached the whole neighborhood yet and there are no prospects for the other neighborhoods. Vila Verde has a street with asphalt and Eldorado has some, but it's very distant and sporadic, it doesn't form an entire route. (Interviewee C - 16/10/2021) .[55]

> When they paved the first street in Maristela, I was still a kid. Then they started paving all around. If you walk there it's like this, the main street and then on the right hand side only one is paved and the other is dirt and if you go straight, then the asphalt is finished. (Interviewee D - 26/10/2021).

In addition to the lack of asphalt based on urban planning, the interviewees pointed out that the only actions taken by the public authorities to solve the problem, which is so disruptive to residents' lives, is the use of earth to level the streets and gravel. Thus:

> Sometimes there's a tractor from the town hall that goes over the road and they throw red earth on it and the people don't like that and I don't know why they keep doing it. Because when it rains, it makes it difficult to pass because it turns into a big muddy mess and cars get stuck and slide. I don't know if they do it to make a point, like 'the town hall is there', but if you ask them they'll see that nobody likes it.
> Mobility on foot and by bicycle or motorcycle is terrible, because there are potholes, dirt and gravel. Sometimes there's so much gravel that you can't stand in front of your house, because if a car comes running past, it flies into you. (Interviewee D).

Images 13 and 14 show the presence of gravel in the streets, which, by the way, has not solved the problems of potholes. In fact, it has made them worse, because even discarded materials (construction rubble) are present in the area, hindering mobility on the streets in this region.

Image 13: Três Lagoas (MS): use of gravel to fill potholes in the streets

[55] Interview conducted on October 16, 2021 with a resident of the Maristela neighborhood.

Image 14: Três Lagoas (MS): use of red earth to lay the street

The interviewees also pointed out that basic sanitation was installed last year, in 2020, and only in the Maristela neighborhood, which, according to the interviewees, is more than 20 years old. According to the interviewees:

146

[Does it have a sanitation system?]

Only in Maristela, since it has been paved, it now has a sewage system and can no longer use cesspits. Maristela is the most structured neighborhood there.
The sewage arrived here in 2020, in Maristela, on my street."
(Interviewee D).

Both Chácara Imperial and Chácara Eldorado, which are more recent neighborhoods, resulting from the process of expansion of the urban network, also highlighted the occurrence of flooding, lack of paving and drainage. In addition, the neighborhoods were considered neglected territory by the public authorities, with difficult access to public transport, a lack of schools, nurseries and health services.

The interviewees reiterated the difficulty of accessing public transport, especially in the Jardim Imperial and Jardim Eldorado neighborhoods:

In the Maristela neighborhood there isn't even a stop, but people already know where it goes, so they already know they can catch it there in the "field". There aren't any bus stops in all the neighborhoods. And when it's time to go back, it's very late, because it's a bus that used to serve this part of the city. When it got to Orestinho, it went all the way up Orestinho, then Vila, then Violeta and then Eldorado. In Eldorado there's no bus stop, you have to get off at specific streets, because it doesn't enter the neighborhood. (Interviewee D).

The interviewees also reported the precarious situation of one of the bus stops, which according to them was a "stick" with "bus" written on it, without a cover and/or a seat, as shown in image 15. The fieldwork showed that in 2020 a proper bus stop was built, with a cover and a seat.

There were also complaints about the quality of public transport, the lack of availability and the long time it takes to make short journeys, such as from the Vila Verde line to UFMS, about 12km.

It's chaos, the buses don't have a stop, for example, in Vila Verde there was no stop, it was just a 'stick', until 2019, buses were written. And so the buses are bad, when it rains it drips, the seats get wet because there's broken glass. To go to college [UFMS] and arrive at 8am, I used to take the bus at 5.30am.

Image 15: Três Lagoas (MS): Vila Verde neighborhood bus stop

Source: Fieldwork, 2021. **Photo:** Baratelli, A. E. S., October 25, 2021.

In addition, these neighborhoods lack basic public facilities for the population. The interviewees reported that there is only one CRAS to serve all the neighborhoods, according to the following statements:

> There's one in Maristela. They built a nursery, then a school and a health center. The CRAS there, if I'm not mistaken, when I worked at the Ana Maria CRAS, the one in Maristela was in Guanabara. Now it seems that Ana Maria is serving Maristela. But Ana Maria is a new CRAS, it's in Vila Verde, and it serves all the neighborhoods. It serves Chácara Imperial, Eldorado, Violeta 1 and 2 and Orestinho and Jardim das Primaveras. (Interviewee D).

In addition to the presence of only one CRAS, the interviewees reported that the same happens with the health center, which has only one unit to serve the entire population of the neighborhoods:

> So, about the health center, there's one in Maristela, I think it serves Violeta I and II, Maristela, Vila Verde, Eldorado and Imperial. One was supposed to open in Eldorado, but it hasn't. It's been delayed for many years, back in 2016. (Interviewee D - 26/10/2021)[56] .

---

[56] Interview conducted on October 26, 2021 with a resident of the Maristela neighborhood.

The post in question, which began to be built in 2016, is behind schedule and has not yet been completed, according to image 16 of the site. The interviewees reported that the area where the post was being built was abandoned and began to accumulate garbage, which led to illness.

> But when there were a lot of outbreaks of leishmaniasis, they started building a health center, it started at the beginning of 2016 and it's still not ready. The town hall abandoned the building, they didn't clean it, then there started to be a lot of outbreaks of leishmaniasis in the neighborhood, so I called them and told them that they had to clean it, that there was an outbreak of the disease and they had to solve it. I think I even reported it to Toninha Campos (the city's radio host). But they cleaned it up like that and the post is still there without being finished." (Interviewee D).

Image 16: Três Lagoas (MS): UBS under construction since 2016 in Jardim Eldorado

Source: Fieldwork, 2021. **Photo:** Baratelli, A. E. S., October 25, 2021.

The neglect of public authorities in this area of the city is compounded in several instances, including access to education. According to the statements, there are only two municipal schools to serve the population and the nearest state school is Luiz Lopes, about 2km to 5km from the neighborhoods analyzed. In addition, mobility for access to school is a contentious issue in the municipality, according to the reports:

> There are two municipal schools here, General Nelson Custodio Oliveira and Marlene de Noronha. The nearest state school is Luiz Lopes, in the Santa Terezinha neighborhood.
> I remember when I was in college, I don't know if it was the state or the municipality that had taken away the school pass from public transportation and there is no specific school bus." (Interviewee E).

149

The interviewees, when asked about the profile of the residents who live in this region, pointed out that it is a poor population. This can be seen in an empirical analysis of the pattern of the houses, as shown in images 17 and 18.

> Low income. I don't know if that's the right word for it, but it's a poor population. A very poor population lives there. One or two have a better situation, but you can see that they've bought the homes of a third party. There are houses that stand out from the rest, because not all of them have walls, because not all of them are finished. Because you can see that sometimes another house is being built, so you can imagine that it's the daughter who lives on the same plot.
>
> From my observation, I can see that these are poorer families. You can tell by the structure of the houses that they are low-income. The houses don't have internal or external plastering and some aren't even walled. This tells me that these are low-income families. The cars are older too (Interviewee D).

Image 17: Três Lagoas (MS): houses with no external plastering

Source: Fieldwork, 2021. **Photo:** Baratelli, A. E. S., October 25, 2021.

Image 18: Houses not yet walled

In addition to the empirical analysis of the housing situation of the population living in this region, more aggravating information was also gathered, such as the situation of food insecurity faced by some families. According to the interviewees, the only action taken by the public authorities to resolve the situation of these families is via the Ana Maria Social Assistance Reference Center (CRAS), shown in image 19.

> Because there are children, like one time a boy was eating desperately, I told him he was going to be sick, then he told me 'auntie, I'm only going to eat tomorrow at school'. They didn't even like going, because they found the games boring, but then their mothers didn't have anyone to leave them with and maybe because they didn't have anything to eat at home. So it's a very needy community there.

> People have been thrown in and there's no infrastructure, the only one that has is CRAS, because even with defects it saves, because the child is taken care of and has the fundamental thing, food. There are no social projects, if there are any for the children it's because the community does something or people from outside take things. The only square there is near Vila Verde and there's a soccer pitch, but it's just the court. (Interviewee D).

Image 19: CRAS Ana Maria

151

Source: Fieldwork, 2021. Photo: Baratelli, A. E. S., October 25, 2021.

The interviews, fieldwork and data collected by the SEBRAE survey reveal severe inconsistencies in the creation of the allotments studied, since they do not follow the urban standards established by the City Statute (2001). According to Chapter I of the City Statute, the basic infrastructure for housing areas is considered to be the preliminary provisions for setting up a subdivision:

> §5° The basic infrastructure of subdivisions is made up of urban equipment for rainwater drainage, public lighting, sanitary sewage, drinking water supply, public and household electricity and roads.
> §Paragraph 6: The basic infrastructure of subdivisions located in housing zones
> declared by law as being of social interest (ZHIS) will consist at least of:
> I - traffic routes;
> II - rainwater runoff;
> III - drinking water supply network; and
> IV- solutions for sanitary sewage and household electricity.

According to the field research, the standards established by the City Statute (2001) were not preserved by the municipality during the development of the allotments and housing estates. According to the interviewees, the population living in this part of the city feels abandoned by the government, with no prospect of improvements to the problems pointed out:

> We were excluded. We used to say we lived in another city. My grandfather used to say "I'm going to Três Lagoas", because it's far

152

away, there's no mobility, the bus is faulty, there's no stop, there's nothing, so for an elderly person to stand there is bad.
In fact, the feeling I get is that it's not going to change any time soon. Because Vila Verde and Maristela are old, because when I moved there they already existed and then other neighborhoods grew. (Interviewee D).

When asked about the presence of pulp companies in the municipality, whose discourse is based on the progress and development of the city, the interviewees' statements reveal the contradictions that occur in practice.

[In these neighborhoods of the city, of exclusion, this generation of employment, this possibility that people believed the companies were going to generate, has arrived].
No, that's a myth. Even because people said they thought it would improve, but it didn't improve at all. When I used to talk to my grandfather, everyone thought it was going to provide jobs, but you'll see that there are only a handful of people who work in this sector. In the pulp sector it's one or two, you see one or two standing on the corner waiting for the bus. (Interviewee D).

According to the bill establishing Três Lagoas as the "national pulp capital", the sector in question had invested large sums of money to improve the lives of the population. However, the interviewees' responses about the actions of these companies in the area studied reveal a lack of knowledge of this fact, as follows:

[Is there action by companies to improve the neighborhoods?]
Not that I know of. I know that at the entrance to the Novo Oeste and Orestinho housing estates, a square was built which, from the plaque, seems to have been supported by the paper mills. I've never seen anything here (Interviewee D - 26/10/2021).

The interviewees pointed out that the distancing of public authorities in these neighborhoods can be justified by the lack of economic interests in the area. They believe that the City Council's attention is focused on the new neighborhoods because there is an interest in selling plots, which is why they invest in urban infrastructure:

I particularly believe that since it's a neighborhood that the population has already structured itself in terms of housing, for example, there's no land to be sold, there aren't any large plots that belong to a single person, they're plots that already have owners. So it doesn't need to be attractive to be sold. So it's different from the new neighborhoods and subdivisions to be sold, where if someone wants to buy, they look to see if there's lighting, asphalt, sewage to see if it's worth buying land in that area. From my point of view, the land is already inhabited. So why make them attractive? (Interviewee D - 26/10/2021).

The idea that investments should only be made if there is a financial return highlights the market nature of the qualification, a situation that favors real estate speculation via "empty" spaces practiced by those who have a monopoly on land in Três Lagoas. The process of speculation should be combated by the public authorities, but it seems that the city council continues to pay special attention to the plots that are being developed. The presence of urban facilities is known to increase the price of plots.

It can also be seen that investments in new allotments are aimed at the most favored groups, because the plots are sold at high prices, as highlighted in the new Santa Helena allotment, where the starting price is R\$92,000. The data from the Participatory Master Plan revealed the presence of real estate speculation, materialized in the high prices of land and the presence of around 30,000 empty plots, as shown:

> With regard to the difficulties of owning a home in the city of Três Lagoas, the group pointed out that the main factor is real estate speculation and the difficulty of financing (bureaucratization). The group concluded by saying that yes, the city has many urban voids, around 30,000 empty lots (source: City Hall). (PDP, 2016, p. 147).

However, even with the significant presence of urban "voids" throughout the city, the process of forming new allotments continues.

It can therefore be concluded that the negligence of public authorities in these neighborhoods is related to a lack of economic interest. Their interests are focused on land where there is the possibility of speculating and earning income. In this way, the rentier nature of the three-lagoon elite and public authorities is evident. In this model of rentier capitalism, whose essence lies in the monopoly of land, the social needs of disadvantaged groups are neglected.

It is also important to note that the processes of real estate speculation and land concentration are not naturalized by the population. The actions of resistance in Três Lagoas, such as the urban occupations, reveal the contradictions of the constant dispute over territory. The first urban occupation in the municipality of Três Lagoas took place in 2000 and was organized by the National Union for Popular Housing Movement, as shown in image 20, where the families demanded the right to housing (BARATELLI; MILANI, 2019).

Image 20: Image of the National Union for Popular Housing Movement in Três Lagoas

Source: BARATELLI; MILANI (2019, p. 91).

The movement was made up of around 144 families who occupied various plots of land in the city to win the right to housing. The movement kept up the arduous struggle for land for 19 years, but it was only in 2019 that construction began on the houses where Jardim das Primaveras is now located. The territorialization of these families in the new subdivision is still problematic, since there is no urban equipment installed. As a result, any activity linked to the public sector has to be carried out in the nearest neighborhood, i.e. the Maristela neighborhood.

In addition to the occupation that formed the Jardim das Primaveras subdivision, others are also present in the city. In 2020, through the Luta Popular movement, the São João urban occupation began[57] , illustrated by image 24. The representative of the occupying group, in a speech to the Três Lagoas City Council, justified the action with data about the significant housing *deficit,* coupled with the fact that many families cannot afford to pay rent and/or buy plots, due to the high price of land. These families are demanding housing based on Articles 182 and 183 of the Federal Constitution, which state that land must have a social function.

Image 21: São João Occupation, Três Lagoas

---

[57] RCN 67. **More than 100 families occupy area in search of their own home**. Available at <https://www.rcn67.com.br/jpnews/tres-lagoas/sem-teto-invadem-area-da-prefeitura-no-bairro-sao-joao/151657/>

Source: *Facebook* images, 2021.

Therefore, the *slogan* given to the municipality, which has created the image of a municipality whose essence is based on agro-industrial development with full employment and citizenship for all, does not match the reality of the population of Tres Lobos.

The official narrative of economic growth resulting from the territorialization of the pulp and paper complex in Três Lagoas is a reality, but this wealth is selective and is not part of the daily life of the dispossessed residents. On the contrary, the presence of these globalized enterprises lays bare the abyss of existing social inequalities, which condemns the municipality's population to live amid the contradictions of the monopoly of land and the expanded reproduction of capital.

FINAL CONSIDERATIONS

The work in question concentrated its efforts on achieving the proposed objective, which was to investigate the realization of the land-capital alliance, as well as its particularities in the face of the territorialization of the eucalyptus-cellulose agribusiness, on the scale of the municipality of Três Lagoas. It therefore set out to understand how this process takes place in the relationship between the countryside and the city, understanding that these two fractions are dialectically complementary in the single capitalist territory.

The Brazilian capitalist state, whose founding premises are patrimonialist (FAORO, 1958), was formed and continues to play the role of defending the interests of the ruling classes. More than that, the state mediates the conflict inherent in the dispute between antagonistic classes, such as between industrial capitalists and landowners. State funding and exemptions have enabled the land-capital alliance, exercised by the dominant classes in Brazil to maintain their privileges (MARTINS, 1994), via conservative modernization (PIRES, RAMOS, 2019).

The reflection on the manifestation of the land-capital alliance, as a backdrop to the private realization of the class reproduction of the Tres-Lagoense elite, was represented by the social practice of the Thomé, Salomão and Prata Tibery families, who belong to the class of landowners in the municipality and the state and continue to diversify their economic activities in order to earn income and profit. And even in the face of economic changes, particularly those brought about by the eucalyptus-pulp sector, they have maintained and expanded their territorial investments.

The territorial formation of Três Lagoas is part of the legacy of the constitution of the state of Mato Grosso and, later, its division to form Mato Grosso do Sul. In both territorial formations, the monopoly of land by oligarchic groups marked the constitution of society and local power. And these days, these practices based on land power are still present in Mato Grosso do Sul, like Governor Reinaldo Azambuja (PSDB), who, according to Castilho (2012), is considered one of the largest landowners in Mato Grosso do Sul and owns 2,500 hectares in Maracaju alone, with declared assets of R$24 million.

The families that represent part of the Tres-Lago elite are well known to the local population. In other words, in the local imagination, they represent a group of successful pioneers who, in the present, want to give back to the municipality some of the wealth they have accumulated. However, it escapes the population's notice that these elites are based on rentier relations of land monopoly. In turn, control of the territory is also the ideological control of society by the ruling class, a situation that allows their dominant thinking to be seen as the thinking of the majority of society, a practice that is essential for them to remain politically and economically privileged.

Because of the monopoly on land and wealth, and the presence of this culture of power, the change in the axis of economic activities resulting from the territorialization of the pulp and paper complex has not substantially altered economic and political relations in the municipality. What's more, wealth is becoming increasingly concentrated because the local dominant groups have diversified their earnings, since the land structure in which these economic changes took place was held captive in a few privileged hands.

This means that although these forestry companies expanded the territorialization of monocultures on a global scale from Três Lagoas, there was no significant change in the distribution of income or in local power relations. On the contrary, landowners used the presence of the territorialization of capital to increase their profits, particularly in the urban areas.

The expansion of eucalyptus monocultures and the territorialization of the cellulose-paper complex have created a growing demand for land, either to lease to the eucalyptus plantations or to set up new urban subdivisions, thus increasing the price of land disproportionately. INCRA's data, which has already been shown in this article, shows how this dynamic has made it more difficult to carry out agrarian reform. Real estate speculation reveals the interests of landowners in increasing the percentage of their land income through land appreciation.

The group of local landowners therefore began to direct their ventures towards the creation of allotments, land sales and the incorporation of hectares into the urban perimeter, intensifying the ways in which they earned income and, above all, increased their wealth. Therefore, the landowning tradition of the Três Lagoense elite has not waned with the expansion of the eucalyptus-cellulose business. On the contrary, in the face of the expansion of the business world, they tried to unite landowners and urban capitalists (land-capital alliance). This is probably the new class movement of contemporary capitalism.

Finally, despite the hegemonic territorial control by landowners and capitalists on different scales, the contradiction is present in every movement of totality carried out by society. Thus, even with the obstacles to social reproduction, the dispossessed continue to resist and lead the class struggle, as has been done by the urban occupations in Três Lagoas and the agroecological peasant recreation in the countryside.

ANNEXES

frente para a Rua A10, possuindo o Lote n. 01 a área de 525,10m²; Lote n. 02 a área de 455,00m²; Lote n. 03 a área de 455,00m²; Lote n. 04 a área de 455,00m²; Lote n. 05 a área de 455,00m²; Lote n. 06 a área de 455,00m²; Lote n. 07 a área de 455,00m²; Lote n. 08 a área de 455,00m²; Lote n. 09 a área de 455,00m²; Lote n. 10 a área de 455,00m²; Lote n. 11 a área de 455,00m²; Lote n. 12 a área de 455,00m²; Lote n. 13 a área de 455,00m²; Lote n. 14 a área de 455,00m²; Lote n. 15 a área de 455,00m²; Lote n. 16 a área de 455,00m²; Lote n. 17 a área de 455,00m²; Lote n. 18 a área de 455,00m²; Lote n. 19 a área de 455,00m²; Lote n. 20 a área de 388,22m²; Lotes n. 21 a 40, com frente para a Rua A11, possuindo o Lote n. 21 a área de 427,05m²; Lote n. 22 a área de 455,00m²; Lote n. 23 a área de 455,00m²; Lote n. 24 a área de 455,00m²; Lote n. 25 a área de 455,00m²; Lote n. 26 a área de 455,00m²; Lote n. 27 a área de 455,00m²; Lote n. 28 a área de 455,00m²; Lote n. 29 a área de 455,00m²; Lote n. 30 a área de 455,00m²; Lote n. 31 a área de 455,00m²; Lote n. 32 a área de 455,00m²; Lote n. 33 a área de 455,00m²; Lote n. 34 a área de 455,00m²; Lote n. 35 a área de 455,00m²; Lote n. 36 a área de 455,00m²; Lote n. 37 a área de 455,00m²; Lote n. 38 a área de 455,00m²; Lote n. 39 a área de 455,00m²; Lote n. 40 a área de 479,97m²; conforme matrículas n. 69.572 à 69.611 (respectivamente).

QUADRA "A9", com a área de 12.687,13m²; constituída por (26) lotes, numerados de 01 a 26; compreendidos entre as Ruas: A01, A07, A09, A10 e AV. A01, sendo os Lotes n. 01 a 12, com frente para a Rua A07, possuindo o Lote n. 01 a área de 734,73m²; Lote n. 02 a área de 478,25m²; Lote n. 03 a área de 478,25m²; Lote n. 04 a área de 478,25m²; Lote n. 05 a área de 478,25m²; Lote n. 06 a área de 478,25m²; Lote n. 07 a área de 478,25m²; Lote n. 08 a área de 478,25m²; Lote n. 09 a área de 471,80m²; Lote n. 10 a área de 455,00m²; Lote n. 11 a área de 455,00m²; Lote n. 12 a área de 450,66m²; Lotes n. 13 a 26, com frente formada por dois segmentos, para a Rua A9 e Rua A10, possuindo o Lote n. 13 a área de 489,49m²; Lote n. 14 a área de 455,00m²; Lote n. 15 a área de 445,00m²; Lote n. 16 a área de 449,26m²; Lote n. 17 a área de 454,73m²; Lote n. 18 a área de 454,73m²; Lote n. 19 a área de 454,73m²; Lote n. 20 a área de 454,73m²; Lote n. 21 a área de 454,73m²; Lote n. 22 a área de 454,73m²; Lote n. 23 a área de 454,73m²; Lote n. 24 a área de 454,73m²; Lote n. 25 a área de 454,73m²; Lote n. 26 a área de 810,87m²; conforme matrículas n. 69.612 à 69.637 (respectivamente).

QUADRA "A10", com a área de 8.548,05m²; constituída por (18) lotes, numerados de 01 a 18; compreendidos entre as Ruas: A01, A06, A07 e AV. A01, sendo os Lotes n. 01 a 08, com frente para a Rua A06, possuindo o Lote n. 01 a área de 530,88m²; Lote n. 02 a área de 490,14m²; Lote n. 03 a área de 490,14m²; Lote n. 04 a área de 490,14m²; Lote n. 05 a área de 495,06m²; Lote n. 06 a área de 455,00m²; Lote n. 07 a área de 455,00m²; Lote n. 08 a área de 450,96m²; Lotes n. 09 a 18, com frente para a Rua A07, possuindo o Lote n. 09 a área de 489,79m²; Lote n. 10 a área de 455,00m²; Lote n. 11 a área de 455,00m²; Lote n. 12 a área de 459,22m²; Lote n. 13 a área de 455,50m²; Lote n. 14 a área de 455,50m²; Lote n. 15 a área de 455,50m²; Lote n. 16 a área de 455,50m²; Lote n. 17 a área de 455,50m²; Lote n. 18 a área de 554,22m²; conforme matrículas n. 69.638 à 69.655 (respectivamente).

QUADRA "A11", com a área de 37.383,16m²; constituída por (15) lotes comerciais, numerados de 01 a 15; compreendidos entre a Rua Vicinal, Terras de Jefferson Jorge Salomão e Outros, Área de Paisagismo 1, e Av. Externa, sendo os Lotes n. 01 a 15, com frente para a Rua Vicinal, possuindo o Lote n. 01 a área de 1.627,24m²; Lote n. 02 a área de 1.178,00m²; Lote n. 03 a área de 1.178,00m²; Lote n. 04 a área de 1.178,00m²; Lote n. 05 a área de 1.178,00m²; Lote n. 06 a área de 1.178,00m²; Lote n. 07 a área de 1.178,00m²; Lote n. 08 a área de 1.178,00m²; Lote n. 09 a área de 1.178,00m²; Lote n. 10 a área de 1.178,00m²; Lote n. 11 a área de 1.178,00m²; Lote n. 12 a área de 1.178,00m²; Lote n. 13 a área de 1.178,00m²; Lote n. 14 a área de 1.182,46m²; Lote n. 15 a área de 20.437,46m²; conforme matrículas n. 69.656 à 69.670 (respectivamente).

ANNEX 1 - REGISTRATION 63.684

IMÓVEL.- Urbano desmembrado da Fazenda Santa Helena, com a área de 48,4000ha. (quarenta e oito hectares e quarenta ares), ou seja, 484.000,375 m², perímetro: 2.831,642m, localizado nesta cidade e comarca de Três Lagoas/MS, dentro das seguintes medidas e confrontações: inicia-se a descrição deste perímetro no vértice EIO P F378, de coordenadas N 7.704.938,2373m e E 428.092,7704m; deste, segue confrontando com Proprietário: Jeferson Jorge Salomão, Matrículas: 47.505 e 60.609, com os seguintes azimutes e distâncias: 300°41'48" e 20,022 m até o vértice EIO M 3923, de coordenadas N 7.704.948,4582m e E 428.075,5542m; 300°42'18" e 935,778 m até o vértice EIO M 3924, de coordenadas N 7.705.426,2852m e E 427.270,8657m; 31°41'30" e 505,710 m até o vértice EIO P F376, de coordenadas N 7.706.866,5877m e E 427.636,6389m; deste, segue confrontando com Proprietários: Claudenor Zopone Junior e outro, Matrículas: 60.265 e 60.263, com os seguintes azimutes e distâncias: 121°04'28" e 970,222 m até o vértice EIO P F377, de coordenadas N 7.705.355,8044m e E 428.367,6305m; deste, segue até o eixo da rodovia que liga o Balneário Municipal à Rodovia BR 158 de Acesso ao Balneário, com os seguintes azimutes e distâncias: 213°21'17" e 499,910 m até o vértice EIO P F378, ponto inicial da descrição deste perímetro. Todas as coordenadas aqui descritas estão georreferenciadas ao Sistema Geodésico Brasileiro, a partir do ponto a BASE, e encontram-se representadas no Sistema UTM, referenciadas ao Meridiano Central nº 51°00' fuso-22, tendo como datum o SIRGAS2000. Todos os azimutes e distâncias, área e perímetro foram calculados no plano de projeção UTM. Memorial descritivo datado de 12 de novembro de 2012, elaborado e assinado pelo Eng. Agrônomo Kennides Martins Batista – CREA: 7.891 – VISTO/MS, ART n. 11408414. Registro Anterior: Matrícula n. 58.440 e 58.442, livro 02, deste Registro Imobiliário. Proprietários: na proporção de 65% (sessenta e cinco por cento) do imóvel objeto desta matrícula, a CLAUDENOR ZOPONE JUNIOR, brasileiro, engenheiro civil, portador da cédula de identidade RG n. 10.347.069-4-SSP/SP, e do CPF/MF n. 067.826.958-04, casado sob o regime da comunhão parcial de bens, na vigência da Lei 6.515/77, com MARCELI JACOB ZOPONE, brasileira, nutricionista; e na proporção de 35% (trinta e cinco por cento) do imóvel objeto desta matrícula a CLAUDIO ZOPONE, brasileiro, engenheiro civil, portador de cédula de identidade RG n. 14.808.391-SSP/SP, e do CPF/MF n. 131.114.538-98, casado sob o regime da comunhão parcial de bens, na vigência da Lei 6.515/77, com FABIANE BUZALAF ZOPONE, brasileira, empresária, todos residentes e domiciliados na Avenida Rodrigues Alves, n. 34-53, Vila Carolina, na cidade de Bauru/SP. Emolumentos: R$ 18,00; FUNJECC 10%: R$ 1,80; FUNJECC 3%: R$ 0,54. Eu, Priscila Kelly da Silva Neto, auxiliar extrajudicial, digitei. Eu, Giovani Gomes Roman, escrevente autorizado, conferi. Dou fé. Três Lagoas/MS, 02 de janeiro de 2013. Oficial/Substituto/Escrevente Autorizado. -

---

R.01/M.63.684.- Protocolo: 177.757 em 29/01/2014.- **Incorporação**.- Pelo requerimento datado de 18 de dezembro de 2013, firmado por Z-Incorporações Imobiliárias Ltda, o qual juntou o Instrumento Particular de Constituição de Sociedade Empresaria, datado de 10 de maio de 2010, registrado na Junta Comercial do Estado de São Paulo, em 21/05/2010, Primeira Alteração de Contrato Social, datado de 25 de setembro de 2013, registrado na Junta Comercial de São Paulo/SP, em 09/10/2013 sob n. 345.358/13-2; Segunda Alteração do Contrato datado de 27 de outubro de 2013, registrado na Junta Comercial do Estado de São Paulo/SP, em 11/11/2013 sob n. 390.429/13-2, bem como a Terceira Alteração do Contrato datado de 22 de novembro de 2013, registrado na Junta Comercial do Estado de São Paulo/SP, em 16/12/2013 sob n. 422.886/13-0, o imóvel objeto da presente matrícula foi transmitido a título de Integralização de Capital a empresa Z-INCORPORAÇÕES IMOBILIÁRIAS LTDA, inscrita no CNPJ/MF sob n. 11.990.848/0001-21, com sede na Rua Francisco de Souza Barbosa, n. 1-80, Sala 01, Vila Monlevade, na cidade de Bauru/SP, no valor de R$ 10.500.000,00 (DEZ MILHÕES E QUINHENTOS MIL REAIS). Apresentaram Certidão Simplificada, emitida em 28 de janeiro de 2014, pela Junta Comercial do Estado de São Paulo; Certidão Negativa de Débitos Relativos às Contribuições Previdenciárias e às de Terceiros n. 002072014-88888848, emitida em 08/01/2014, pela Secretaria da Receita

Federal do Brasil, com validade até 07/07/2014 em nome de Z-Incorporações Imobiliárias Ltda; Certidão Conjunta de Débitos Relativos aos Tributos Federais e à Dívida Ativa da União, em nome de Z-Incorporações Imobiliárias Ltda, emitida em 18/12/2013, pela Secretaria da Receita Federal do Brasil, com validade até 16/06/2014, código de controle da certidão: 505A.23A8.EDA5.4E86; Certificado de Regularidade do FGTS – CRF, em nome de Z-Incorporações Imobiliárias Ltda, emitida pela Caixa Econômica Federal, com validade de 09/01/2014 à 07/02/2014, certificação número: 2014010908444731615050. Não incidência do pagamento do ITBI, conforme Processo sob protocolo n. 23.116 de 19 de dezembro de 2013, artigo 51, inciso I, da Lei 1.087/91 (CTM), pela Secretaria Municipal de Finanças e Planejamento Núcleo de Julgamento e Consultas, em 20 de dezembro de 2013, assinado pelo julgador Éderson Felix da Silva – Matrícula 2-16269, Certidão Negativa de Débito Municipal, protocolado sob n. 1316/2014, expedido em 24/01/2014, pela Prefeitura Municipal local. Imóvel cadastrado na Prefeitura sob n. 7.71.000.0000.00132. Emolumentos: R$ 914,00; FUNJECC 10%: R$ 91,40; FUNJECC 3%: R$ 27,42. FUNADEP R$ 18,40. Selo digital n. AGU 02345-380 (este selo poderá ser conferido e autenticado no site: www.tjms.jus.br/corregedoria/selos/pesquisaSelo.php). Eu, Dayane Ramires de Oliveira Facholi, auxiliar extrajudicial, digitei. Eu, Simone de Lima Moreira, auxiliar extrajudicial, conferi. Dou fé. Três Lagoas/MS, 28 de fevereiro de 2014. Oficial/Substituto/Escrevente Autorizado. -

R.02/M.63.684.- Protocolo:- 181.369 em 25/06/2014. - Loteamento. - O imóvel objeto desta matrícula foi totalmente LOTEADO, conforme planta e memoriais descritivos aprovados pela Prefeitura Municipal desta cidade, nos termos do Alvará de aprovação n. 649/2014, expedido em 21/07/2014, devidamente assinado pelo Eng. Rodrigo Pelho Rizzo, Diretor de Dep. de Fiscalização de Obras, CREA/VISTO-MS 14.095, conforme consta do processo arquivado neste cartório. O Loteamento, de propriedade de Z-INCORPORAÇÕES IMOBILIÁRIAS LTDA., já qualificada, com a denominação de "LOTEAMENTO FECHADO RESIDENCIAL VILLA DUMONT", contendo a área total de 484.000,38m², subdividido da seguinte forma:

I – ÁREAS PÚBLICAS:-

SISTEMA VIÁRIO: com a área de 138.894,11m², correspondente a (28,70%).

Leito carroçável/passeios: com a área de 132.153,41m².
Canalização viária: com a área de 6.740,70m².

BACIA DE INFILTRAÇÃO: com a área de 20.054,91m², correspondente a (4,14%)

TALUDE BACIA DE INFILTRAÇÃO: com a área de 6.858,72, correspondente a (1,41%)

ÁREAS VERDES: com a área de 8.782,07m², correspondente a (1,82%).

Área Verde 1: com a área de 2.191,55m².

Área Verde 2: com a área de 1.342,43m².

Área Verde 3: com a área de 2.584,69m².

163

Área Verde 4: com a área de 302,79m².

Área Verde 5: com a área de 302,79m².

TOTAL DE ÁREAS PÚBLICAS: 174.589,81m², correspondente a (36,07%).

II - ÁREA DOS LOTES:-

RESIDENCIAL UNIFAMILIAR: 507 (quinhentos e sete), com a área de 221.474,88m², correspondente a (45,76%).
LOTES COMERCIAIS: 16 (dezesseis), com a área de 38.933,04m², correspondente a (8,04%).
LOTES CLUBE: 02 (dois), com a área de 6.572,41m², correspondente a (1,36%).
LOTES DE LAZER: 03 (três), com a área de 15.518,01m², correspondente a (3,21%).
LOTES DE PORTARIA: 02 (dois), com a área 690,16m², correspondente a (0,14%).
LOTES DE SERVIÇO: 02 (dois), com a área de 896,30m², correspondente a (0,19%).
ÁREA DE PAISAGISMO: 06 (seis) lotes, com a área de 25.325,77m², correspondente a (5,23%). **TOTAL DE LOTES**: 538 (quinhentos e trinta e oito), com a área de 309.410,57m², correspondente a (63,93%); subdividida em 23 (vinte e três) — Quadras, assim designadas:

QUADRA "A1", com a área de 4.377,96m²; constituída por (08) lotes, numerados de 01 a 08; compreendidos entre as Ruas: A06, A07, Avenida: A01 e Área de Paisagismo 3, sendo os **Lotes n. 01 a 03**, com frente para a Rua A06, possuindo o **Lote n. 01**, a área de 548,51m², **Lote n. 02** a área de 603,42m² e o **Lote n.03** a área de 518,37m²; sendo os **Lotes n. 04 a 08**, com frente para a Rua A07, possuindo o **Lote n. 04** a área de 554,64m², **Lote n. 05** a área de 455,50m², **Lote n. 06** a área de 455,50m², **Lote n. 07** a área de 455,50m² e o **Lote n. 08** a área de 786,52m²; **conforme matrículas n. 69.448 à 69.455 (respectivamente)**.

QUADRA "A2", com a área de 7.342,80m²; constituída por (14) lotes, numerados de 01 a 14; compreendidos entre as Ruas: A07, A08, A09 e Avenida A01, sendo os **Lotes n. 01 a 06**, com frente para a Rua A07, possuindo o **Lote n. 01** a área de 716,75m²; **Lote n.02** a área de 478,25m²; **Lote n. 03** a área de 478,25m²; **Lote n. 04** a área de 478,25m²; **Lote n. 05** a área de 478,25m² e o **Lote n. 06** a área de 711,46m²; e os **Lotes n. 07 a 14**, com frente para a Rua A09, possuindo o **Lote n. 07** a área de 628,63m²; **Lote n. 08** a área de 454,73m²; **Lote n. 09** a área de 454,73m²; **Lote n. 10** a área de 454,73m²; **Lote n. 11** a área de 454,73m²; **Lote n. 12** a área de 454,73m²; **Lote n. 13** a área de 454,73m² e o **Lote n. 14** a área de 644,58m²; **conforme matrículas n. 69.456 à 69.469 (respectivamente)**.

QUADRA "A3", com a área de 11.577,80m²; constituída por (24) lotes, numerados de 01 a 24; compreendidos entre as Ruas: A03, A04, A08 e 10A, sendo os **Lotes n. 01 a 12**, com frente para a Rua A03, possuindo o **Lote n. 01** a área de 525,32m²; **Lote n.02** a área de 455,00m²; **Lote n. 03** a área de 455,00m²; **Lote n. 04** a área de 455,00m²; **Lote n. 05** a área de 455,00m²; **Lote n. 06** a área de 455,00m²; **Lote n. 07** a área de 455,00m²; **Lote n. 08** a área de 455,00m²; **Lote n. 09** a área de 455,00m²; **Lote n. 10** a área de 455,00m²; **Lote n. 11** a área de 455,00m² e o **Lote n. 12** a área de 857,69m²; e os **Lotes n. 13 a 24**, com frente para a Rua A04, possuindo o **Lote n. 13** a área de 614,14m²; **Lote n. 14** a área de 455,00m²; **Lote n. 15** a área de 455,00m²; **Lote n. 16** a área de 455,00m²; **Lote n. 17** a área de 455,00m²; **Lote n.18** a área de 455,00m²; **Lote n. 19** a área de 455,00m²; **Lote n. 20** a área de

164

455,00m²; Lote n. 21 a área de 455,00m²; Lote n. 22 a área de 455,00m²; Lote n. 23 a área de 455,00m²; e Lote n. 24 a área de 480,65m²; conforme matrículas n. 69.470 à 69.493 (respectivamente).

QUADRA "A4", com a área de 10.776,04m²; constituída por (22) lotes, numerados de 01 a 22; compreendidos entre as Ruas: A04, A05, A06 e 10A, sendo os Lotes n. 01 a 11, com frente para a Rua A04, possuindo o Lote n. 01 a área de 525,32m²; Lote n.02 a área de 455,00m²; Lote n.03 a área de 455,00m²; Lote n.04 a área de 455,00m²; Lote n.05 a área de 455,00m²; Lote n.06 a área de 455,00m²; Lote n.07 a área de 455,00m²; Lote n.08 a área de 455,00m²; Lote n.09 a área de 455,00m²; Lote n.10 a área de 455,00m²; Lote n.11 a área de 688,19m²; sendo os Lotes n.12 a 22, com frente para a Rua A05, possuindo o Lote n.12 a área de 891,88m²; Lote n.13 a área de 455,00m²; Lote n.14 a área de 455,00m²; Lote n.15 a área de 455,00m²; Lote n.16 a área de 455,00m²; Lote n.17 a área de 455,00m²; Lote n.18 a área de 455,00m²; Lote n.20 a área de 455,00m²; Lote n.21 a área de 455,00m²; Lote n.22 a área de 480,65m²; conforme matrículas n. 69.494 à 69.515 (respectivamente).

QUADRA "A5", com a área de 9.300,14m²; constituída por (20) lotes, numerados de 01 a 20; compreendidos entre as Ruas: A04, A05, A10 e A12, sendo os Lotes n. 01 a 10, com frente para a Rua A04, possuindo o Lote n. 01 a área de 484,97m²; Lote n. 02 a área de 455,00m²; Lote n. 03 a área de 455,00m²; Lote n. 04 a área de 455,00m²; Lote n. 05 a área de 455,00m²; Lote n. 06 a área de 455,00m²; Lote n. 07 a área de 455,00m²; Lote n. 08 a área de 455,00m²; Lote n. 09 a área de 455,00m² e Lote n. 10 a área de 525,10m²; Lotes n. 11 a 20, com frente para a Rua A05, possuindo o Lote n. 11 a área de 479,97m²; Lote n. 12 a área de 455,00m²; Lote n. 13 a área de 455,00m²; Lote n. 14 a área de 455,00m²; Lote n. 15 a área de 455,00m²; Lote n. 16 a área de 455,00m²; Lote n. 17 a área de 455,00m²; Lote n. 18 a área de 455,00m²; Lote n. 19 a área de 455,00m²; Lote n. 20 a área de 530,10m²; conforme matrículas n. 69.516 à 69.535 (respectivamente).

QUADRA "A6", com a área de 9.300,14m²; constituída por (20) lotes, numerados de 01 a 20; compreendidos entre as Ruas: A03, A04, A10 e A12, sendo os Lotes n. 01 a 10, com frente para a Rua A03, possuindo o Lote n. 01 a área de 484,97m²; Lote n. 02 a área de 455,00m²; Lote n. 03 a área de 455,00m²; Lote n. 04 a área de 455,00m²; Lote n. 05 a área de 455,00m²; Lote n. 06 a área de 455,00m²; Lote n. 07 a área de 455,00m²; Lote n. 08 a área de 455,00m²; Lote n. 09 a área de 455,00m²; Lote n. 10 a área de 525,10m²; Lotes n. 11 a 20, com frente para a Rua A04, possuindo o Lote n. 11 a área de 479,97m²; Lote n. 12 a área de 455,00m²; Lote n. 13 a área de 455,00m²; Lote n. 14 a área de 455,00m²; Lote n. 15 a área de 455,00m²; Lote n. 16 a área de 455,00m²; Lote n. 17 a área de 455,00m²; Lote n. 18 a área de 455,00m²; Lote n. 19 a área de 455,00m²; Lote n. 20 a área de 530,10m²; conforme matrículas n. 69.536 à 69.555 (respectivamente).

QUADRA "A7", com a área de 7.342,13m²; constituída por (16) lotes, numerados de 01 a 16; compreendidos entre as Ruas: A02, A03, A11 e A12, sendo os Lotes n. 01 a 08, com frente para a Rua A11, possuindo o Lote n. 01 a área de 525,10m²; Lote n. 02 a área de 455,00m²; Lote n. 03 a área de 455,00m²; Lote n. 04 a área de 455,00m²; Lote n. 05 a área de 455,00m²; Lote n. 06 a área de 455,00m²; Lote n. 07 a área de 455,00m²; Lote n. 08 a área de 438,53m²; Lotes n. 09 a 16, com frente para a Rua A12, possuindo o Lote n. 09 a área de 438,53m²; Lote n. 10 a área de 455,00m²; Lote n. 11 a área de 455,00m²; Lote n. 12 a área de 455,00m²; Lote n. 13 a área de 455,00m²; Lote n. 14 a área de 455,00m²; Lote n. 15 a área de 455,00m²; Lote n. 16 a área de 479,97m²; conforme matrículas n. 69.556 à 69.571 (respectivamente).

QUADRA "A8", com a área de 18.200,34m²; constituída por (40) lotes, numerados de 01 a 40; compreendidos entre as Ruas: A01, A03, A10 e A11, sendo os Lotes n. 01 a 20, com

QUADRA "A12", com a área de 1.546,88m²; constituída por (01) lote comercial; compreendido entre terras de Claudenor Zopone Junior e Outros, Av. Externa e Sistema Viário, sendo o **Lote n. 01**, com frente para Av. Externa e Sistema Viário, possuindo a área de 1.546,88m²; conforme matrícula n. 69.671.

QUADRA "B1", com a área de 10.952,73m²; constituída por (26) lotes, numerados de 01 a 26; compreendidos entre a Rua: B01, terras de Claudenor Zopone Junior e outros e o Lote Clube 2 e Área de Paisagismo 6; sendo os **Lotes n. 01 a 26**, com frente para a Rua B01, possuindo o **Lote n. 01** a área de 844,50m²; **Lote n. 02** a área de 390,00m²; **Lote n. 03** a área de 390,00m²; **Lote n. 04** a área de 390,00m²; **Lote n. 05** a área de 390,00m²; **Lote n. 06** a área de 390,00m²; **Lote n. 07** a área de 390,00m²; **Lote n. 08** a área de 390,00m²; **Lote n. 09** a área de 390,00m²; **Lote n. 10** a área de 390,00m²; **Lote n. 11** a área de 390,00m²; **Lote n. 12** a área de 390,00m²; **Lote n. 13** a área de 390,00m²; **Lote n. 14** a área de 390,00m²; **Lote n. 15** a área de 390,00m²; **Lote n. 16** a área de 390,00m²; **Lote n. 17** a área de 390,00m²; **Lote n. 18** a área de 390,00m²; **Lote n. 19** a área de 390,00m²; **Lote n. 20** a área de 390,00m²; **Lote n. 21** a área de 390,00m²; **Lote n. 22** a área de 390,00m²; **Lote n. 23** a área de 390,00m²; **Lote n. 24** a área de 390,00m²; **Lote n. 25** a área de 390,46m²; **Lote n. 26** a área de 747,77m²; conforme matrículas n. 69.672 a 69.697 (respectivamente).

QUADRA "B2", com a área de 8.159,64m²; constituída por (19) lotes, numerados de 01 a 19; compreendidos entre as Ruas: B01, B11, B13 e AV. B01, sendo os **Lotes n. 01 a 08**, com frente para a Rua B13, possuindo o **Lote n. 01** a área de 641,89m²; **Lote n. 02** a área de 450,56m²; **Lote n. 03** a área de 452,39m²; **Lote n. 04** a área de 452,39m²; **Lote n. 05** a área de 452,39m²; **Lote n. 06** a área de 416,12m²; **Lote n. 07** a área de 390,00m²; **Lote n. 08** a área de 410,40m²; **Lotes n. 09 a 19**, com frente para a Rua B11, possuindo o **Lote n. 09** a área de 399,89m²; **Lote n. 10** a área de 390,00m²; **Lote n. 11** a área de 374,10m²; **Lote n. 12** a área de 392,97m²; **Lote n. 13** a área de 400,80m²; **Lote n. 14** a área de 400,80m²; **Lote n. 15** a área de 400,80m²; **Lote n. 16** a área de 400,80m²; **Lote n. 17** a área de 400,80m²; **Lote n. 18** a área de 400,80m²; **Lote n. 19** a área de 531,74m²; conforme matrículas n. 69.698 a 69.716 (respectivamente).

QUADRA "B3", com a área de 12.191,87m²; constituída por (29) lotes, numerados de 01 a 29; compreendidos entre as Ruas: B01, B08, B11 e AV. B01, sendo os **Lotes n. 01 a 13**, com frente para a Rua B11, possuindo o **Lote n. 01** a área de 589,85m²; **Lote n. 02** a área de 425,09m²; **Lote n. 03** a área de 425,09m²; **Lote n. 04** a área de 425,09m²; **Lote n. 05** a área de 425,09m²; **Lote n. 06** a área de 425,09m²; **Lote n. 07** a área de 425,09m²; **Lote n. 08** a área de 425,09m²; **Lote n. 09** a área de 425,09m²; **Lote n. 10** a área de 425,09m²; **Lote n. 11** a área de 409,78m²; **Lote n. 12** a área de 390,00m²; **Lote n. 13** a área de 469,70m²; **Lotes n. 14 a 29**, com frente para a Rua B08, possuindo o **Lote n. 14** a área de 464,27m²; **Lote n. 15** a área de 390,00m²; **Lote n. 16** a área de 389,28m²; **Lote n. 17** a área de 384,75m²; **Lote n. 18** a área de 384,75m²; **Lote n. 19** a área de 384,75m²; **Lote n. 20** a área de 384,75m²; **Lote n. 21** a área de 384,75m²; **Lote n. 22** a área de 384,75m²; **Lote n. 23** a área de 384,75m²; **Lote n. 24** a área de 384,75m²; **Lote n. 25** a área de 384,75m²; **Lote n. 26** a área de 384,75m²; **Lote n. 27** a área de 384,75m²; **Lote n. 28** a área de 384,75m²; **Lote n. 29** a área de 645,98m²; conforme matrículas n. 69.717 a 69.745 (respectivamente).

QUADRA "B4", com a área de 18.217,46m²; constituída por (46) lotes, numerados de 01 a 46; compreendidos entre as Ruas: B01, B02, B06 e B07, sendo os **Lotes n. 01 a 23**, com frente para a Rua B07, possuindo o **Lote n. 01** a área de 461,70m²; **Lote n. 02** a área de 390,00m²; **Lote n. 03** a área de 390,00m²; **Lote n. 04** a área de 390,00m²; **Lote n. 05** a área de 390,00m²; **Lote n. 06** a área de 390,00m²; **Lote n. 07** a área de 390,00m²; **Lote n. 08** a área de 390,00m²; **Lote n. 09** a área de 390,00m²; **Lote n. 10** a área de 390,00m²; **Lote n. 11** a área de 390,00m²; **Lote n. 12** a área de 390,00m²; **Lote n. 13** a área de 390,00m²; **Lote n.**

14 a área de 390,00m²; Lote n. 15 a área de 390,00m²; Lote n. 16 a área de 390,00m²; Lote n. 17 a área de 390,00m²; Lote n. 18 a área de 390,00m²; Lote n. 19 a área de 390,00m²; Lote n. 20 a área de 390,00m²; Lote n. 21 a área de 390,00m²; Lote n. 22 a área de 390,00m²; Lote n. 23 a área de 453,88m²; Lotes n. 24 a 46, com frente para a Rua B06, possuindo o Lote n. 24 a área de 443,38m²; Lote n. 25 a área de 390,00m²; Lote n. 26 a área de 390,00m²; Lote n. 27 a área de 390,00m²; Lote n. 28 a área de 390,00m²; Lote n. 29 a área de 390,00m²; Lote n. 30 a área de 390,00m²; Lote n. 31 a área de 390,00m²; Lote n. 32 a área de 390,00m²; Lote n. 33 a área de 390,00m²; Lote n. 34 a área de 390,00m²; Lote n. 35 a área de 390,00m²; Lote n. 36 a área de 390,00m²; Lote n. 37 a área de 390,00m²; Lote n. 38 a área de 390,00m²; Lote n. 39 a área de 390,00m²; Lote n. 40 a área de 390,00m²; Lote n. 41 a área de 390,00m²; Lote n. 42 a área de 390,00m²; Lote n. 43 a área de 390,00m²; Lote n. 44 a área de 390,00m²; Lote n. 45 a área de 390,00m²; Lote n. 46 a área de 478,50m²; conforme matrículas n. 69.746 à 69.791 (respectivamente).

QUADRA "B5", com a área de 18.260,60m²; constituída por (46) lotes, numerados de 01 a 46; compreendidos entre as Ruas: B01, B02, B05 e B06, sendo os Lotes n. 01 a 23, com frente para a Rua B06, possuindo o Lote n. 01 a área de 460,08m²; Lote n. 02 a área de 390,00m²; Lote n. 03 a área de 390,00m²; Lote n. 04 a área de 390,00m²; Lote n. 05 a área de 390,00m²; Lote n. 06 a área de 390,00m²; Lote n. 07 a área de 390,00m²; Lote n. 08 a área de 390,00m²; Lote n. 09 a área de 390,00m²; Lote n. 10 a área de 390,00m²; Lote n. 11 a área de 390,00m²; Lote n. 12 a área de 390,00m²; Lote n. 13 a área de 390,00m²; Lote n. 14 a área de 390,00m²; Lote n. 15 a área de 390,00m²; Lote n. 16 a área de 390,00m²; Lote n. 17 a área de 390,00m²; Lote n. 18 a área de 390,00m²; Lote n. 19 a área de 390,00m²; Lote n. 20 a área de 390,00m²; Lote n. 21 a área de 390,00m²; Lote n. 22 a área de 390,00m²; Lote n. 23 a área de 472,07m²; Lotes n. 24 a 46, com frente para a Rua B05, possuindo o Lote n. 24 a área de 461,57m²; Lote n. 25 a área de 390,00m²; Lote n. 26 a área de 390,00m²; Lote n. 27 a área de 390,00m²; Lote n. 28 a área de 390,00m²; Lote n. 29 a área de 390,00m²; Lote n. 30 a área de 390,00m²; Lote n. 31 a área de 390,00m²; Lote n. 32 a área de 390,00m²; Lote n. 33 a área de 390,00m²; Lote n. 34 a área de 390,00m²; Lote n. 35 a área de 390,00m²; Lote n. 36 a área de 390,00m²; Lote n. 37 a área de 390,00m²; Lote n. 38 a área de 390,00m²; Lote n. 39 a área de 390,00m²; Lote n. 40 a área de 390,00m²; Lote n. 41 a área de 390,00m²; Lote n. 42 a área de 390,00m²; Lote n. 43 a área de 390,00m²; Lote n. 44 a área de 390,00m²; Lote n. 45 a área de 390,00m²; Lote n. 46 a área de 476,88m²; conforme matrículas n. 69.792 à 69.837 (respectivamente).

QUADRA "B6", com a área de 9.291,81m²; constituída por (24) lotes, numerados de 01 a 24; compreendidos entre as Ruas: B02, B03, B05 e B07, sendo os Lotes n. 01 a 12, com frente para a Rua B02, possuindo o Lote n. 01 a área de 381,70m²; Lote n. 02 a área de 390,00m²; Lote n. 03 a área de 390,00m²; Lote n. 04 a área de 390,00m²; Lote n. 05 a área de 390,00m²; Lote n. 06 a área de 390,00m²; Lote n. 07 a área de 390,00m²; Lote n. 08 a área de 390,00m²; Lote n. 09 a área de 390,00m²; Lote n. 10 a área de 390,00m²; Lote n. 11 a área de 390,00m²; Lote n. 12 a área de 364,21m²; Lotes n. 13 a 24, com frente para a Rua B03, possuindo o Lote n. 13 a área de 381,00m²; Lote n. 14 a área de 390,00m²; Lote n. 15 a área de 390,00m²; Lote n. 16 a área de 390,00m²; Lote n. 17 a área de 390,00m²; Lote n. 18 a área de 390,00m²; Lote n. 19 a área de 390,00m²; Lote n. 20 a área de 390,00m²; Lote n. 21 a área de 390,00m²; Lote n. 22 a área de 390,00m²; Lote n. 23 a área de 390,00m²; Lote n. 24 a área de 364,90m²; conforme matrículas n. 69.838 à 69.861 (respectivamente).

QUADRA "B7", com a área de 9.291,81m²; constituída por (24) lotes, numerados de 01 a 24; compreendidos entre as Ruas: B03, B04, B05 e B07, sendo os Lotes n. 01 a 12, com frente para a Rua B03, possuindo o Lote n. 01 a área de 381,70m²; Lote n. 02 a área de 390,00m²; Lote n. 03 a área de 390,00m²; Lote n. 04 a área de 390,00m²; Lote n. 05 a área de 390,00m²; Lote n. 06 a área de 390,00m²; Lote n. 07 a área de 390,00m²; Lote n. 08 a área de 390,00m²; Lote n. 09 a área de 390,00m²; Lote n. 10 a área de 390,00m²; Lote n. 11 área de 390,00m²;

a área de 390,00m²; Lote n. 12 a área de 364,21m²; Lotes n. 13 a 24, com frente para a Rua B04, possuindo o Lote n. 13 a área de 381,00m²; Lote n. 14 a área de 390,00m²; Lote n. 15 a área de 390,00m²; Lote n. 16 a área de 390,00m²; Lote n. 17 a área de 390,00m²; Lote n. 18 a área de 390,00m²; Lote n. 19 a área de 390,00m²; Lote n. 20 a área de 390,00m²; Lote n. 21 a área de 390,00m²; Lote n. 22 a área de 390,00m²; Lote n. 23 a área de 390,00m²; Lote n. 24 a área de 364,90m²; conforme matrículas n. 69.862 à 69.885 (respectivamente).

QUADRA "B8", com a área de 11.619,02m²; constituída por (29) lotes, numerados de 01 a 29; compreendidos entre as Ruas: B03, B04, B07 e B12, sendo os Lotes n. 01 a 14, com frente para a Rua B03, possuindo o Lote n. 01 a área de 396,71m²; Lote n. 02 a área de 390,00m²; Lote n. 03 a área de 390,00m²; Lote n. 04 a área de 390,00m²; Lote n. 05 a área de 390,00m²; Lote n. 06 a área de 390,00m²; Lote n. 07 a área de 390,00m²; Lote n. 08 a área de 390,00m²; Lote n. 09 a área de 390,00m²; Lote n. 10 a área de 390,00m²; Lote n. 11 a área de 390,00m²; Lote n. 12 a área de 390,00m²; Lote n. 13 a área de 390,00m²; Lote n. 14 a área de 622,33m²; Lotes n. 15 a 29, com frente para a Rua B04, possuindo o Lote n. 15 a área de 436,48m²; Lote n. 16 a área de 390,00m²; Lote n. 17 a área de 390,00m²; Lote n. 18 a área de 390,00m²; Lote n. 19 a área de 390,00m²; Lote n. 20 a área de 390,00m²; Lote n. 21 a área de 390,00m²; Lote n. 22 a área de 390,00m²; Lote n. 23 a área de 390,00m²; Lote n. 24 a área de 390,00m²; Lote n. 25 a área de 390,00m²; Lote n. 26 a área de 390,00m²; Lote n. 27 a área de 390,00m²; Lote n. 28 a área de 390,00m²; Lote n. 29 a área de 413,50m²; conforme matrículas n. 69.886 à 69.914 (respectivamente).

QUADRA "B9", com a área de 12.182,39m²; constituída por (29) lotes, numerados de 01 a 29; compreendidos entre as Ruas: B02, B03, B07 e B12, sendo os Lotes n. 01 a 15, com frente para a Rua B02, possuindo o Lote n. 01 a área de 396,71m²; Lote n. 02 a área de 390,00m²; Lote n. 03 a área de 390,00m²; Lote n. 04 a área de 390,00m²; Lote n. 05 a área de 390,00m²; Lote n. 06 a área de 390,00m²; Lote n. 07 a área de 390,00m²; Lote n. 08 a área de 390,00m²; Lote n. 09 a área de 390,00m²; Lote n. 10 a área de 390,00m²; Lote n. 11 a área de 390,00m²; Lote n. 12 a área de 390,00m²; Lote n. 13 a área de 390,00m²; Lote n. 14 a área de 390,00m²; Lote n. 15 a área de 755,54m²; Lotes n. 16 a 29, com frente para a Rua B03, possuindo o Lote n. 16 a área de 866,64m²; Lote n. 17 a área de 390,00m²; Lote n. 18 a área de 390,00m²; Lote n. 19 a área de 390,00m²; Lote n. 20 a área de 390,00m²; Lote n. 21 a área de 390,00m²; Lote n. 22 a área de 390,00m²; Lote n. 23 a área de 390,00m²; Lote n. 24 a área de 390,00m²; Lote n. 25 a área de 390,00m²; Lote n. 26 a área de 390,00m²; Lote n. 27 a área de 390,00m²; Lote n. 28 a área de 390,00m²; Lote n. 29 a área de 413,50m²; conforme matrículas n. 69.915 à 69.943 (respectivamente).

QUADRA "B10", com a área de 7.412,43m²; constituída por (17) lotes, numerados de 01 a 17; compreendidos entre as Ruas: B08, B11, B12, e Av. B01, sendo os Lotes n. 01 a 07, com frente para a Rua B11, possuindo o Lote n. 01 a área de 626,40m²; Lote n. 02 a área de 425,37m²; Lote n. 03 a área de 425,37m²; Lote n. 04 a área de 425,37m²; Lote n. 05 a área de 425,37m²; Lote n. 06 a área de 425,37m²; Lote n. 07 a área de 625,03m²; Lotes n. 08 a 17, com frente para a Rua B08, possuindo o Lote n. 08 a área de 467,42m²; Lote n. 09 a área de 384,75m²; Lote n. 10 a área de 384,75m²; Lote n. 11 a área de 384,75m²; Lote n. 12 a área de 384,75m²; Lote n. 13 a área de 384,75m²; Lote n. 14 a área de 384,75m²; Lote n. 15 a área de 384,75m²; Lote n. 16 a área de 384,75m²; Lote n. 17 a área de 488,73m²; conforme matrículas n. 69.944 à 69.960 (respectivamente).

QUADRA "B11", com a área de 4.472,79m²; constituída por (10) lotes, numerados de 01 a 10; compreendidos entre as Ruas B11, B13, Av. B01, e Área de Paisagismo 5, sendo os Lotes n. 01 a 04, com frente para a Rua B13, possuindo o Lote n. 01 a área de 463,25m²; Lote n. 02 a área de 451,28m²; Lote n. 03 a área de 373,55m²; Lote n. 04 a área de 432,87m²; n. 05 a 10, com frente para a Rua B11, possuindo o Lote n. 05 a área de

590,24m²; **Lote n. 06** a área de 400,80m²; **Lote n. 07** a área de 400,80m²; **Lote n. 08** a área de 400,80m²; **Lote n. 09** a área de 400,80m²; **Lote n. 10** a área de 558,40m²; **conforme matrículas n. 69.961 à 69.970 (respectivamente)**.

Será doada externamente a gleba uma área de 38.720,00m², equivalente a 8% da área total da gleba, destinada a área pública para equipamentos comunitários.
Será doada externamente a gleba uma área de 31.873,93m², equivalente a complementação da área verde (6,58%) da área total da gleba.

As áreas Verdes e Institucionais passaram a integrar ao domínio do Município de Três Lagoas/MS, conforme Artigo 22, da Lei 6.766 de 19/12/1979. Emolumentos R$ 9.684,00; FUNJECC 10% R$ 968,40; FUNJECC 3% R$ 290,52; FUNADEP 6% R$ 581,04; FUNDE-PGE 4% R$ 387,36. Selo digital n. AHX91576-637 (este selo poderá ser conferido e autenticado no site: www.tjms.jus.br/corregedoria/selos/pesquisaSelo.php). Eu, Fernanda Andrade Moura de Almeida, auxiliar extrajudicial, digitei. Eu, Giovani Gomes Roman, escrevente autorizado, conferi. **Data:** Três Lagoas/MS, 08 de agosto de 2014. (Ass) Jacqueline Yamaguti Ueda, Escrevente autorizada. Dou fé. Três Lagoas/MS, 20 de novembro de 2014. Oficial/Substituto/Escrevente autorizado _________________________

Av.03/M.63.684.- Prenotação:- 184.143 em 21/10/2014.- Pelo requerimento datado de 15 de outubro de 2014, firmado pela proprietária, **Z- Incorporações Imobiliárias Ltda.**, já qualificada, faz-se a presente averbação para constar a abertura de matrícula das áreas de uso comum do **"LOTEAMENTO FECHADO RESIDENCIAL VILLA DUMONT"**, assim designadas:

"Lote Lazer 1 — Fase A", com a área de 6.734,74m²; constituído por (01) lote, compreendido entre as Ruas: A03, A09, A10, A13 e A14, conforme matrícula 71.163;

"Lote Lazer 3 — Fase B", com a área de 8.374,08m²; constituído por (01) lote, compreendido entre as Ruas: A02, A07, A08, A09 e A10, conforme matrícula 71.164;

"Lote Serviço 1 — Fase A", com a área de 449,31m²; constituído por (01) lote, compreendido entre a Rua A06; Avenida A01; Área Verde 01; conforme matrícula 71.165;

"Lote Serviço 2 — Fase B", com a área de 446,99m²; constituído por (01) lote, compreendido entre a Rua B13; Avenida B01 e Área de Paisagismo 5; conforme matrícula 71.166;

"Lote Portaria 1 — Fase A", com a área de 345,08m²; constituído por (01) lote, com frente para a Avenida A01; conforme matrícula 71.167;

"Lote Portaria 2 — Fase B", com a área de 345,08m²; constituído por (01) lote, com frente para a Avenida B01; conforme matrícula 71.168;

"Lote Clube 1 — Fase A", com a área de 2.715,83m²; constituído por (01), compreendido entre a Rua A06; Avenida Externa; Lote de Serviço 01 e Área verde 1; conforme matrícula 71.169;

"Lote Clube 2 — Fase B", com a área de 3.856,58m²; constituído por (01), compreendido entre a Rua B13; Rua Externa 02; Lote 26 da Quadra B1 e Lote de Lazer 2; conforme matrícula 71.170, todas do livro 02, deste RI. Eu, Fernanda Andrade Moura de Almeida, auxiliar extrajudicial, digitei. Eu, Giovani Gomes Roman, escrevente autorizado, conferi. **Data:** Três Lagoas/MS, 19 de novembro de 2014. Dou fé. Três Lagoas/MS, 20 de novembro de 2014. Oficial/Substituto/Escrevente autorizado _________________________

Av.04/M.63.684.- Em virtude de falha na impressão dos atos praticados nesta matrícula, foi informado ao Juiz Diretor do Foro desta Comarca de Três Lagoas/MS, Dr. Márcio Rogério Alves, através do Ofício n. 1.188/2014, datado de 20/11/2014, a reimpressão desta ficha; conforme determinado em Correição realizada na data de 09/10/2007, pelo MM. Juiz Diretor do Foro – Dr. Renato Antônio de Liberali. Dou fé. Três Lagoas/MS. 20 de novembro de 2014. Oficial/Substituto/Escrevente Autorizado.-

Av.05/M.63.684.- Prenotação:- 195.933 em 15/04/2016.- Pelo requerimento datado de 15 de abril de 2016, firmado pela proprietária, **Z- Incorporações Imobiliárias Ltda**, já qualificada, faz-se a presente averbação para constar a abertura de matrícula das áreas de uso comum do **"LOTEAMENTO FECHADO RESIDENCIAL VILLA DUMONT"**, assim designadas:

"ÁREA DE PAISAGISMO 2", com a área de 1.342,44m²; constituído por (01) lote, compreendido entre as Avenidas B01 e A01 e Rua Externa, conforme matrícula 77.965;

"ÁREA DE PAISAGISMO 4", com a área de 3.362,89m²; constituído por (01) lote, compreendido entre as Ruas: A08, B12, e Áreas de Paisagismo 1, 3, 4 e 5, conforme matrícula 77.966, ambas do livro 02, deste RI. Eu, Antonio Magusso Neto, auxiliar extrajudicial, digitei. Eu, Giovani Gomes Roman, escrevente autorizado, conferi. Três Lagoas/MS, 19 de novembro de 2014. Dou fé. Três Lagoas/MS, 09 de maio de 2016. Oficial/Substituto/Escrevente autorizado

CERTIFICO que a presente fotocópia confere com a matrícula original de n. 63684 e que, nos termos do disposto artigo 19, § 1º da Lei 6.015/1973, tem valor de certidão. O referido é verdade e dou fé. Três Lagoas, MS, em 22 de dezembro de 2021. SELO nº AFU68876-883-NOR.

Oficial do Registro / Substituto / Escrevente

Acesse o site http://www.tjms.jus.br/corregedoria/selos/pesquisaselos.php para visualizar a

ANNEX 2 - REGISTRATION 62.714

Imóvel: Rural destacado da Fazenda Vó Ruthy com a área de 48,4000ha (quarenta e oito hectares e quarenta ares), situada neste município e comarca de Três Lagoas/MS, dentro da seguinte descrição: inicia-se a descrição deste perímetro no vértice M13; Tipo de divisa cerca; deste, segue confrontando com o Corredor Publico, com azimute 120°59'34" e distancia de 222,84m até o vértice M01; Tipo de divisa de cerca; deste, segue confrontando com a área remanescente pertencente a Orestes Prata Tibery Junior, com os seguintes azimutes e distancias: 176°53'55" e 39,78m até o vértice M06; 171°17'33" e 291,84m até o vértice M05; 210°27'34" e 20,20m até o vértice M04; 121°01'51" e 742,67m até o vértice M03; Tipo de divisa de cerca; deste, segue confrontando com a área pertencente a Orestes Prata Tibery Junior, com o azimute de 210°29'21" e 176,93m até o vértice M10; Tipo de divisa cerca; deste, segue confrontando com a faixa de domínio da Ferrovia, com os seguintes azimutes e distancias: 263°38'11" e 18,98m até o vértice AGH-M-2847; 259°42'20" e 25,00m até o vértice AGH-M-2848; 254°37'34" e 24,48m até o vértice AGH-M-8084; 252°48'46" e 24,78m até o vértice AGH-M8084; 252°48'46" e 24,78m até o vértice AGH-M8085; 251°07'39" e 41,65m até o vértice AGH-M8086; 227°19'14" e 72,49m até o vértice M11; Tipo de divisa cerca, deste, segue confrontando com a Area 2 (remanescente), pertencente a Orestes Prata Tibery Junior, com os seguintes azimutes e distancias: 300°53'09" e 1.060,34m até o vértice M12; 30°53'09" e 622,68m até o vértice M13; ponto inicial da descrição deste perímetro. Memorial Descritivo, elaborado e assinado pelo Engenheiro Civil Pedro Donizete Bortolote, CREA 0600994590, ART n. 11344137. **Registro anterior:** Matrícula n. 58.575, livro 02 deste Registro Imobiliário. **Proprietário: ORESTES PRATA TIBERY JUNIOR**, brasileiro, pecuarista, portador do RG n. 524.096-SSP/MS e do CPF/MF n. 008.024.681-87, casado sob o regime de separação obrigatória de bens, na vigência da Lei 6.515/77, nos termos do artigo 1.641, parágrafo único, II, do Código Civil, com **ELLEN PERBONI MARTINS PRATA TIBERY**, brasileira, hoteleira, portadora do RG n. 24.266.630-9-SSP/SP e do CPF/MF n. 792.503.351-72, residentes e domiciliados na Rua João Gonçalves de Oliveira, n. 810, Vila Nova, nesta cidade de Três Lagoas/MS. Emolumentos: R$ 18,00; FUNJECC 10%: R$ 1,80; FUNJECC 3%: R$ 0,54. Eu, Simone de Lima Moreira, auxiliar extrajudicial, digitei. Eu, Douglas Rodrigo Damasceno Fernandes, auxiliar extrajudicial, conferi. Dou fé. Três Lagoas/MS, 03 de setembro de 2012. Oficial/Substituto/Escrevente Autorizado.

R.01/M.62.714.-Protocolo:- 164.468 em 07/08/2012. -**Venda e compra**. - Pela escritura pública de venda e compra lavrada no livro n. 764, f. 338, em 17 de abril de 2012, pelo Registro Civil das Pessoas Naturais e Tabelião de Notas, Distrito de Engenheiro Schmidt, comarca de São Jose do Rio Preto/SP, o proprietário **ORESTES PRATA TIBERY JUNIOR**, com anuência de sua esposa Ellen Perboni Martins Prata Tibery, já qualificado, **vendeu** o imóvel objeto da presente matrícula a **VIIV EMPREENDIMENTOS IMOBILIARIOS – SPE TRÊS LAGOAS LTDA**, inscrita no CNPJ/MF n. 15.322.434/0001-85, com sede na Rua Bahia, n. 630, parte 2, Centro, na cidade de Catanduva/SP, presentada por seu procurador especial, Luciano Sanches Fernandes, brasileiro, casado, empresário, portador do RG n. 16.393.274-8-SSP/SP e do CPF/MF n. 098.197.408-27, residente e domiciliado na Rua Bareirinha, n. 225, Jardim dos Coqueiros, na cidade de Catanduva/SP, pelo preço de R$ 7.000.000,00 (SETE MILHÕES DE REAIS), que confessa e declara já haver recebido anteriormente e integralmente dos outorgados comprador, em forma de depósitos feitos na conta corrente numero 16400-3, agencia 0208-9, do Banco do Brasil S/A, de titularidade do outorgante Orestes Prata Tibery Junior, da seguinte forma: (A) R$ 500.000,00 (quinhentos mil reais), em 19 de dezembro de 2011, por meio das seguintes transferências bancarias: (I) R$ 55.555,55 (cinquenta e cinco mil quinhentos e cinquenta e cinco reais e cinquenta e cinco centavos), da conta corrente n. 00136-0, agencia 4004 do Banco Itaú S/A, de titularidade de Breno Fernandes Dias; (II) o valor de R$ R$ 55.555,55 (cinquenta e cinco mil quinhentos e cinquenta e cinco reais e cinquenta e cinco centavos), da conta corrente n. 00134-5, agencia 4004 do Banco Itaú S/A, de titularidade de Caio Fernandes Dias, (III) o valor de R$ 55.555,55 (cinquenta e cinco mil quinhentos e cinquenta e cinco reais e cinquenta e cinco centavos), da conta corrente numero 002256-6, agencia 4004 do Banco Itaú S/A, de titularidade de Marcela Fernandes Dias, (IV) o valor de R$ 166.666,67 (cento e sessenta e seis mil seiscentos e sessenta e seis reais e sessenta e sete centavos), da conta corrente numero 00140-2, agencia 4004 do Banco Itaú S/A, de titularidade de Luciano Sanches Fernandes e (v) o valor de R$ 166.666,68 (cento e sessenta e seis mil seiscentos e sessenta e seis reais e sessenta e oito centavos) da conta corrente numero 0020052-2, agencia 3635-8 do Banco do Bradesco S/A, de titularidade de Andrea Sanches Fernandes; (B) R$ 3.000.000,00 (três milhões de reais), em 15 de fevereiro de 2012, por meio de transferência bancaria da conta corrente numero 10011-00, agencia 2.042, do Banco Bradesco S/A, de titularidade de VIIV Empreendimentos Imobiliários S/A; e (C) R$ 3.500.000,00 (três milhões e quinhentos mil reais), através de deposito bancário da conta corrente n. 11.020-5, do Banco Bradesco S/A, agencia 2042-7, de titularidade da outorgada compradora; cuja importância dá plena, geral, rasa e irrevogável quitação, pagos e satisfeitos, para nunca mais repetir. Apresentaram o pagamento do ITBI no valor de R$ 140.000,00 sobre 2% da avaliação do imóvel em R$ 7.000.000,00, conforme guia n. 80948979, expedida em 08/05/2012, pela Prefeitura Municipal desta cidade de Três Lagoas/MS. Certidão Negativa de Débitos Relativos ao Imposto sobre a Propriedade Territorial Rural emitida em 29/03/2012 pela Secretaria da Receita Federal do Brasil, com validade até 25/09/2012, código de controle da certidão: 9ABA.EF4.AB78.CA84, onde consta o NIRF 2.486.889-2, Fazenda Vó Ruth, município de Três Lagoas/MS, área 510,8ha, em nome de Orestes Prata Tibery Junior. CCIR-2006/2007/2008/2009 onde consta o código do imóvel 912.034.014.397-1, área total 416,1166ha, mód.rural 40,0000, n. mód.rurais 8,25, mód.fiscal 35,0000, n. mód.fiscais 11,8891, FMP 2,0000, área registrada 441,7487ha, classificação fundiária: média propriedade produtiva, Fazenda Vó Ruth, município Três Lagoas/MS, em nome de Orestes Prata Tibery Junior, brasileiro, código da pessoa: 00.488.294-6, Certidão Negativa de Débito expedida pelo IBAMA; *fica ciente a compradora que deverá regularizar a "RESERVA LEGAL", nos termos do provimento n. 15 de 24/06/09 da Corregedoria Geral de Justiça de Mato Grosso do Sul.* Emolumentos: R$ 2.481,00; FUNJECC 10%: R$ 248,10; FUNJECC 3%: R$ 74,43. Selo digital n. ADL 65689-207. Eu, Simone de Lima Moreira, auxiliar extrajudicial, digitei. Eu, Douglas Rodrigo Damasceno Fernandes, auxiliar extrajudicial, conferi. Dou fé. Três Lagoas/MS, 03 de setembro de 2012. Oficial/Substituto/Escrevente Autorizado.

Av.02/M.62.714. - Protocolo: - 175.467 em 24/10/2013. - **Descaracterização do imóvel.** - Pelo requerimento datado de 23 de outubro de 2013, a proprietária **VIIV EMPREENDIMENTOS IMOBILIARIOS – SPE TRÊS LAGOAS LTDA**, já qualificada, devidamente representada por seu sócio administrador: Túlio Soubhia Ribeiro, brasileiro, solteiro, maior, administrador de empresas, portador da cédula de identidade RG n. 32.920.205-4-SSP-SP, inscrito no CPF/MF n. 310.407.348-12, residente e domiciliado na Rua Bahia, n. 55, na cidade de Catanduva/SP, requer a presente averbação para constar que de acordo com Declaração emitida em 17 de julho de 2013, devidamente assinada pelo Diretor de Departamento de Administração Tributária da Prefeitura Municipal local, Divino Teodoro dos Santos, e pelo Secretário Municipal de Finanças, Receita e Controle, Gilmar Meneguzzo, certifica que o imóvel objeto desta matrícula encontra-se situada dentro do Perímetro Urbano do Município, conforme Lei Municipal n. 2.236, de 26/12/2007, em atendimento ao disposto nos artigos 3º e 53º da Lei n. 6.766/79, Lei 9.785/99, enquadrando-se nos dispositivos do artigo 32, parágrafo 1º da Lei 5.172/66 e nas disposições da OS.INCRA/DC/Nº11/76, bem como Parecer Técnico expedido pelo INCRA, conforme Ofício n. 1.745/2013/GAB/F datado de 09 de outubro de 2013, devidamente assinado pelo Sr. Celso Menezes de Souza, Superintendente Regional Substituto, ficando portanto o imóvel objeto da presente matrícula **descaracterizado** de rural para imóvel **URBANO**. Emolumentos: R$ 34,00; FUNJECC 10% R$ 3,40; FUNJECC 3% R$ 1,02. Selo digital n. AGC 69525-401 (este selo poderá ser conferido e autenticado no site: www.tjms.jus.br/corregedoria/selos/pesquisaSelo.php). Eu, Simone de Lima Moreira, auxiliar extrajudicial, digitei. Eu, Lucas Vinicius de Oliveira Arruda, auxiliar extrajudicial, conferi. Dou

16. Três Lagoas/MS, 01 de novembro de 2013. Oficial/Substituto/Escrevente Autorizado. -

R.09/M.62.714.- Prenotação:- 186.428 em 11/12/2014. - **Loteamento.** - O imóvel objeto desta matrícula foi totalmente LOTEADO, conforme planta e memoriais descritivos aprovados pela Secretaria de Obras da Prefeitura Municipal desta cidade, nos termos do Alvará de aprovação n. 914/2013, expedido em 15/09/2014, devidamente assinado pelo Eng. Rodrigo Pelho Rizzo, Secretário de Obras, CREA/VISTO-MS 14.095, conforme consta do processo arquivado neste cartório. O Loteamento, de propriedade de **VIIV EMPREENDIMENTOS IMOBILIÁRIOS - SPE TRÊS LAGOAS LTDA**, já qualificada, com a denominação de "**LOTEAMENTO RESIDENCIAL ORESTES PRATA TIBERY JUNIOR**", contendo a área total de 484.000,00m², subdividido da seguinte forma:

I – TOTAL DE ÁREAS PÚBLICAS:- (222.363,26m²) correspondente a (45,94%).

SISTEMA VIÁRIO: com a área de 151.966,73m².

ÁREAS VERDES: com a área de 39.243,10m².

FAIXA DE SEGURANÇA ELEKTRO: com a área de 31.153,43m².

II – TOTAL DE ÁREAS DOS LOTES:- (261.636,74m²) correspondente a (54,06%)

LOTES RESIDENCIAIS: 490 (quatrocentos e noventa) lotes, com a área de 180.497,88m².

LOTES COMERCIAIS: 38 (trinta e oito) lotes, com a área de 14.956,85m².
LOTES MISTOS: 168 (cento e sessenta e oito) lotes, com a área de 66.182,01m², subdividida em 31 (trinta e uma) quadras assim designadas:

QUADRA "01", com a área de 7.246,27m², constituída por 20 (vinte) lotes, numerados de 01 a 20, compreendidos entre as Ruas: Rua Mollana, Rua Rio Sucuriú, Rua Rio Paraná e Avenida Aroeira, sendo os **Lotes n. 01 a 05**, com frente para a Rua Mollana, possuindo o **Lote n. 01** área de 361,18m², **Lote n. 02**, área de 367,02m², **Lote n. 03**, área de 367,55m², **Lote n. 04**, área de 368,07m², **Lote n. 05**, área de 363,19m², os **Lotes n. 06 a 10**, com frente para a Rua Rio Sucuriú, possuindo os **Lotes n. 06 a 10** a área de 360,00m², cada lote; os **Lotes n. 11 a 15**, com frente para a Rua Rio Paraná, possuindo o **Lote n. 11** a área de 360,63m², **Lotes n. 12 a 14** com a área de 366,00m², cada lote, **Lote n. 15** com a área de 360,63m², e os **Lotes n. 16 a 20**, com frente para a Avenida Aroeira, possuindo os **Lotes n. 16 a 20** a área de 360,00m² cada lote; **conforme matrículas n. 72.433 a 72.452 (respectivamente)**.

QUADRA "02", com a área de 3.722,16m², constituída por 09 (nove) lotes, numerados de 01 a 09, compreendidos entre as Ruas: Rua Rio Paraná, Rua Rio Sucuriú, Avenida Rio Amambai e Avenida Aroeira, sendo os **Lotes n. 01 a 05**, com frente para a Rua Rio Paraná, possuindo o **Lote n. 01** a área de 381,70m², **Lotes n. 02 a 04** a área de 392,86m², cada lote, **Lote n. 05** a área de 387,51m², os **Lotes n. 06 a 08**, com frente para a Rua Rio Sucuriú, possuindo o **Lote n. 06** área de 408,00m², **Lote n. 07** área de 428,25m², **Lote n. 08** área de 488,09m², e o **Lote n. 09**, com frente para a Avenida Rio Amambai, possuindo o **Lote n. 09** área de 449,97m², **conforme matrículas n. 72.453 a 72.461 (respectivamente)**.

QUADRA "03", com a área de 7.262,43m², constituída por 20 (vinte) lotes, numerados de 01 a 20, compreendidos entre as Ruas: Rua Mollana, Rua Rio Piquiri, Rua Rio Paraná e Rua Rio Sucuriú, sendo os **Lotes n. 01 a 05**, com frente para a Rua Mollana, possuindo o **Lote n.**

01 área de 364,41m², **Lote n. 02**, área de 370,26m², **Lote n. 03**, área de 370,78m², **Lote n. 04**, área de 371,30m², **Lote n. 05**, área de 366,42m², os **Lotes n. 06 a 10**, com frente para a Rua Rio Piquiri, possuindo os **Lotes n. 06 a 10** a área de 360,00m² cada lote; os **Lotes n. 11 a 15**, com frente para a Rua Rio Paraná, possuindo o **Lote n. 11** a área de 360,63m², **Lotes n. 12 a 14** com a área de 366,00m², cada lote, **Lote n. 15** com a área de 360,63m², e os **Lotes n. 16 a 20**, com frente para a Rua Rio Sucuriú, possuindo os **Lotes n. 16 a 20** a área de 360,00m² cada lote; **conforme matrículas n. 72.462 a 72.481 (respectivamente)**.

QUADRA "04", com a área de 10.114,04m², constituída por 27 (vinte e sete) lotes, numerados de 01 a 27, compreendidos entre as Ruas: Rua Rio Sucuriú, Rua Rio Paraná, Rua Rio Piquiri e Rua Rio Negro, sendo os **Lotes n. 01 a 05**, com frente para a Rua Rio Paraná, possuindo o **Lote n. 01** área de 387,51m², **Lotes n. 02 a 04**, área de 392,88m², cada lote, **Lote n. 05**, área de 387,51m², os **Lotes n. 06 a 16**, com frente para a Rua Rio Piquiri, possuindo os **Lotes n. 06 a 15** a área de 360,00m² cada lote; **Lote n. 16**, área de 498,75m², os **Lotes n. 17 e 18**, com frente para a Rua Rio Negro, possuindo o **Lote n. 17** a área de 361,73m², **Lote n. 18** com a área de 361,67m², e os **Lotes n. 19 a 27**, com frente para a Rua Rio Sucuriú, possuindo o **Lote n. 19** a área de 405,16m²; **Lote n. 20** a área de 413,07m²; os **Lotes n. 21 a 27**, área de 360,00m², cada lote; **conforme matrículas n. 72.482 a 72.508 (respectivamente)**.

QUADRA "05", com a área de 7.279,90m², constituída por 20 (vinte) lotes, numerados de 01 a 20, compreendidos entre as Ruas: Rua Rio Piquiri, Rua Rio Paraná, Rua Rio Taquari e Rua Mollana, sendo os **Lotes n. 01 a 05**, com frente para a Rua Mollana, possuindo o **Lote n. 01** área de 361,06m², **Lote n. 02**, área de 366,91m², **Lote n. 03**, área de 367,43m², **Lote n. 04**, área de 367,95m², **Lote n. 05**, área de 363,07m², os **Lotes n. 06 a 10**, com frente para a Rua Rio Taquari, possuindo os **Lotes n. 06 a 10** a área de 360,00m² cada lote; os **Lotes n. 11 a 15**, com frente para a Rua Rio Paraná, possuindo o **Lote n. 11** a área de 367,48m², **Lotes n. 12 a 14** com a área de 372,84m², cada lote, **Lote n. 15**, área de 367,48m², e os **Lotes n. 16 a 20**, com frente para a Rua Rio Piquiri, possuindo os **Lotes n. 16 a 20** a área de 360,00m², cada lote; **conforme matrículas n. 72.509 a 72.528 (respectivamente)**.

QUADRA "06", com a área de 11.107,32m², constituída por 30 (trinta) lotes, numerados de 01 a 30, compreendidos entre as Ruas: Rua Rio Piquiri, Rua Rio Negro, Rua Rio Taquari e Rua Rio Paraná, sendo os **Lotes n. 01 a 05**, com frente para a Rua Rio Paraná, possuindo o **Lote n. 01** área de 387,51m², **Lotes n. 02 a 04**, área de 392,88m², cada lote, **Lote n. 05**, área de 387,51m², os **Lotes n. 06 a 15**, com frente para a Rua Rio Taquari, possuindo os **Lotes n. 06 a 15** a área de 360,00m² cada lote; os **Lotes n. 16 a 20**, com frente para a Rua Rio Negro, possuindo o **Lote n. 16** a área de 387,51m², **Lotes n. 17 a 19** com a área de 392,88m², cada lote, **Lote n. 20**, área de 387,51m²; e os **Lotes n. 21 a 30**, com frente para a Rua Rio Piquiri, possuindo os **Lotes n. 21 a 30** a área de 360,00m², cada lote; **conforme matrículas n. 72.529 a 72.558 (respectivamente)**.

QUADRA "07", com a área de 7.296,06m², constituída por 20 (vinte) lotes, numerados de 01 a 20, compreendidos entre as Ruas: Rua Rio Taquari, Rua Rio Paraná, Rua Mollana e Rua Rio do Peixe, sendo os **Lotes n. 01 a 05**, com frente para a Rua Mollana, possuindo o **Lote n. 01** área de 364,29m², **Lote n. 02**, área de 370,14m², **Lote n. 03**, área de 370,66m², **Lote n. 04**, área de 371,19m², **Lote n. 05**, área de 366,30m², os **Lotes n. 06 a 10**, com frente para a Rua Rio do Peixe, possuindo os **Lotes n. 06 a 10** a área de 360,00m² cada lote; os **Lotes n. 11 a 15**, com frente para a Rua Rio Paraná, possuindo o **Lote n. 11** a área de 367,48m², **Lotes n. 12 a 14** com a área de 372,84m², cada lote, **Lote n. 15**, área de 367,48m²; e os **Lotes n. 16 a 20**, com frente para a Rua Rio Taquari, possuindo os **Lotes n. 16 a 20** a área de 360,00m², cada lote; **conforme matrículas n. 72.559 a 72.578 (respectivamente)**.

QUADRA "08", com a área de 11.107,32m², constituída por 30 (trinta) lotes, numerados de 01 a 30, compreendidos entre as Ruas: Rua Rio Taquari, Rua Rio Paraná, Rua Rio Negro e

Rua Rio do Peixe, sendo os **Lotes n. 01 a 05**, com frente para a Rua Rio Paraná, possuindo o **Lote n. 01** área de 387,51m², **Lotes n. 02 a 04**, área de 392,88m², cada lote, **Lote n. 05**, área de 387,51m², os **Lotes n. 06 a 15**, com frente para a Rua Rio do Peixe, possuindo os **Lotes n. 06 a 15** a área de 360,00m², cada lote; os **Lotes n. 16 a 20**, com frente para a Rua Rio Negro, possuindo o **Lote n. 16** a área de 387,51m², **Lotes n. 17 a 19** com a área de 392,88m², cada lote, **Lote n. 20**, área de 387,51m²; e os **Lotes n. 21 a 30**, com frente para a Rua Rio Taquari, possuindo os **Lotes n. 21 a 30** a área de 360,00m², cada lote; **conforme matrículas n. 72.679 a 72.608 (respectivamente)**.

QUADRA "09", com a área de 7.313,53m², constituída por 20 (vinte) lotes, numerados de 01 a 20, compreendidos entre as Ruas: Rua Rio Paraná, Rua Rio do Peixe, Rua Mollana, Rua Rio Jauru, sendo os **Lotes n. 01 a 05**, com frente para a Rua Mollana, possuindo o **Lote n. 01** área de 367,79m², **Lote n. 02**, área de 373,63m², **Lote n. 03**, área de 374,16m², **Lote n. 04**, área de 374,68m², **Lote n. 05**, área de 369,79m², os **Lotes n. 06 a 10**, com frente para a Rua Rio Jauru, possuindo os **Lotes n. 06 a 10** a área de 360,00m², cada lote; os **Lotes n. 11 a 15**, com frente para a Rua Rio Paraná, possuindo o **Lote n. 11** a área de 367,48m², **Lotes n. 12 a 14** com a área de 372,84m², cada lote, **Lote n. 15**, área de 367,48m²; e os **Lotes n. 16 a 20**, com frente para a Rua Rio do Peixe, possuindo os **Lotes n. 16 a 20** a área de 360,00m², cada lote; **conforme matrículas n. 72.609 a 72.628 (respectivamente)**.

QUADRA "10", com a área de 11.107,32m², constituída por 30 (trinta) lotes, numerados de 01 a 30, compreendidos entre as Ruas: Rua Rio do Peixe, Rua Rio Negro, Rua Rio Jauru, Rua Rio Paraná, sendo os **Lotes n. 01 a 05**, com frente para a Rua Rio Paraná, possuindo o **Lote n. 01** área de 387,51m², **Lotes n. 02 a 04**, área de 392,88m², cada lote, **Lote n. 05**, área de 387,51m², os **Lotes n. 06 a 15**, com frente para a Rua Rio Jauru, possuindo os **Lotes n. 06 a 15** a área de 360,00m², cada lote; os **Lotes n. 16 a 20**, com frente para a Rua Rio Negro, possuindo o **Lote n. 16** a área de 387,51m², **Lotes n. 17 a 19** com a área de 392,88m², cada lote, **Lote n. 20**, área de 387,51m²; e os **Lotes n. 21 a 30**, com frente para a Rua Rio do Peixe, possuindo os **Lotes n. 21 a 30** a área de 360,00m², cada lote; **conforme matrículas n. 72.629 a 72.658 (respectivamente)**.

QUADRA "11", com a área de 7.329,68m², constituída por 20 (vinte) lotes, numerados de 01 a 20, compreendidos entre as Ruas: Rua Rio Jauru, Rua Rio Paraná, Rua Rio Aquidauana, Rua Mollana, sendo os **Lotes n. 01 a 05**, com frente para a Rua Mollana, possuindo o **Lote n. 01** área de 371,02m², **Lote n. 02**, área de 376,86m², **Lote n. 03**, área de 377,39m², **Lote n. 04**, área de 377,91m², **Lote n. 05**, área de 377,02m², os **Lotes n. 06 a 10**, com frente para a Rua Rio Aquidauana, possuindo os **Lotes n. 06 a 10** a área de 360,00m², cada lote; os **Lotes n. 11 a 15**, com frente para a Rua Rio Paraná, possuindo o **Lote n. 11** a área de 367,48m², **Lotes n. 12 a 14** com a área de 372,84m², cada lote, **Lote n. 15**, área de 367,48m²; e os **Lotes n. 16 a 20**, com frente para a Rua Rio Jauru, possuindo os **Lotes n. 16 a 20** a área de 360,00m², cada lote; **conforme matrículas n. 72.659 a 72.678 (respectivamente)**.

QUADRA "12", com a área de 11.041,10m², constituída por 29 (vinte e nove) lotes, numerados de 01 a 29, compreendidos entre as Ruas: Rua Rio Jauru, Rua Rio Negro, Rua Rio Aquidauana, Rua Rio Paraná, sendo os **Lotes n. 01 a 05**, com frente para a Rua Rio Paraná, possuindo o **Lote n. 01** área de 387,51m², **Lotes n. 02 a 04**, área de 392,88m², cada lote, **Lote n. 05**, área de 387,51m², os **Lotes n. 06 a 15**, com frente para a Rua Rio Aquidauana, possuindo os **Lotes n. 06 a 15** a área de 360,00m², cada lote; os **Lotes n. 16 a 19**, com frente para a Rua Rio Negro, possuindo o **Lote n. 16** a área de 714,17m², **Lotes n. 17 e 18** com a área de 392,88m², cada lote, **Lote n. 19**, área de 387,51m²; e os **Lotes n. 20 a 29**, com frente para a Rua Rio Jauru, possuindo os **Lotes n. 20 a 29** a área de 360,00m², cada lote; **conforme matrículas n. 72.679 a 72.707 (respectivamente)**.

QUADRA "13", com a área de 7.347,16m², constituída por 20 (vinte) lotes, numerados de 01 a 20, compreendidos entre as Ruas: Rua Rio Aquidauana, Rua Rio Paraná, Rua Rio

Taquaruçu, Rua Moliana, sendo os **Lotes n. 01 a 05**, com frente para a Rua Moliana, possuindo o **Lote n. 01** área de 374,51m², **Lote n. 02**, área de 380,36m², **Lote n. 03**, área de 380,88m², **Lote n. 04**, área de 381,41m², **Lote n. 05**, área de 376,52m², os **Lotes n. 06 a 10**, com frente para a Rua Rio Taquaruçu, possuindo os **Lotes n. 06 a 10** a área de 360,00m², cada lote; os **Lotes n. 11 a 15**, com frente para a Rua Rio Paraná, possuindo o **Lote n. 11** a área de 367,48m², **Lotes n. 12 a 14** com a área de 372,84m², cada lote, **Lote n. 15**, área de 367,48m²; e os **Lotes n. 16 a 20**, com frente para a Rua Rio Aquidauana, possuindo os Lotes n. 16 a 20 a área de 360,00m², cada lote; **conforme matrículas n. 72.708 a 72.727 (respectivamente)**.

QUADRA "14", com a área de 9.056,91m², constituída por 01 (um) lote, destinado **ÁREA VERDE 04**, compreendido entre as Ruas: Avenida Rio Miranda, Rua Rio Aquidauana, Rua Rio Paraná, **conforme matrícula n. 72.728**

QUADRA "15", com a área de 14.357,64m², constituída por 01 (um) lote, destinado **ÁREA VERDE 03**, compreendido entre as Ruas: Rua Rio Taquaruçu, Rua Rio Paraná, Avenida Rio Miranda, Rua Moliana, **conforme matrícula n. 72.729**

QUADRA "16", com a área de 13.494,08m², constituída por 01 (um) lote, destinado **ÁREA VERDE 05**, compreendido entre as Ruas: Rua Rio São Domingos, Rua Rio Negro, Avenida Rio Miranda, **conforme matrícula n. 72.730**

QUADRA "17", com a área de 11.107,32m², constituída por 30 (trinta) lotes, numerados de 01 a 30, compreendidos entre as Ruas: Rua Rio São Domingos, Rua Rio Negro, Rua Rio São Lourenço, Rua Rio Paraná, sendo os **Lotes n. 01 a 05**, com frente para a Rua Rio Paraná, possuindo o **Lote n. 01** área de 387,51m², **Lotes n. 02 a 04**, área de 392,88m², cada lote, **Lote n. 05**, área de 387,51m², os **Lotes n. 06 a 16**, com frente para a Rua Rio São Lourenço, possuindo os Lotes n. 06 a 16 a área de 360,00m², cada lote; os **Lotes n. 16 a 20**, com frente para a Rua Rio Negro, possuindo o **Lote n. 16** a área de 387,51m², **Lotes n. 17 a 19** com a área de 392,88m², cada lote, **Lote n. 20**, área de 387,51m²; e os **Lotes n. 21 a 30**, com frente para a Rua Rio São Domingos, possuindo os **Lotes n. 21 a 30** a área de 360,00m², cada lote; **conforme matrículas n. 72.731 a 72.760 (respectivamente)**.

QUADRA "18", com a área de 10.323,15m², constituída por 24 (vinte e quatro) lotes, numerados de 01 a 24, compreendidos entre as Ruas: Avenida Rio Miranda, Rua Rio Paraná, Rua Rio Aporé, sendo os **Lotes n. 01 a 09**, com frente para a Rua Rio Aporé, possuindo o **Lote n. 01** área de 371,24m², **Lote n. 02**, área de 375,42m², **Lotes n. 03 a 06**, área de 360,00m², cada lote, **Lotes n. 07 a 09**, área de 444,00m², cada lote, os **Lotes n. 10 a 18**, com frente para a Rua Rio Paraná, possuindo o **Lote n. 10** a área de 384,63m², **Lotes n. 11 e 12**, área de 360,00m², cada lote; **Lotes n. 13 e 14**, área de 504,00m², cada lote, **Lotes n. 15 a 17**, área de 360,00m², cada lote, **Lote n. 18**, área de 662,66m²; os **Lotes n. 19 a 24**, com frente para a Avenida Rio Miranda, possuindo o **Lote n. 19** a área de 491,22m², **Lote n. 20** a área de 658,14m², **Lote n. 21**, área de 537,00m²; **Lote n. 22**, área de 499,86m²; **Lote n. 23**, área de 378,78m²; **Lote n. 24**, área de 394,20m²; **conforme matrículas n. 72.761 a 72.784 (respectivamente)**.

QUADRA "19", com a área de 11.107,32m², constituída por 30 (trinta) lotes, numerados de 01 a 30, compreendidos entre as Ruas: Rua Rio São Lourenço, Rua Rio Negro, Rua Rio Aporé, Rua Rio Paraná, sendo os **Lotes n. 01 a 05**, com frente para a Rua Rio Paraná, possuindo o **Lote n. 01** área de 387,51m², **Lotes n. 02 a 04**, área de 392,88m², cada lote, **Lote n. 05** área de 387,51m², os **Lotes n. 06 a 15**, com frente para a Rua Rio Aporé, possuindo os **Lotes n. 06 a 15** a área de 360,00m², cada lote; os **Lotes n. 16 a 20**, com frente para a Rua Rio Negro, possuindo o **Lote n. 16** a área de 387,51m², **Lotes n. 17 a 19** a área de 392,88m², cada lote, **Lote n. 20**, área de 387,51m²; e os **Lotes n. 21 a 30**, com frente para a Rua Rio São Lourenço, possuindo os Lotes n. 21 a 30 a área de 360,00m², cada lote; conforme matrículas n. 72.785 a 72.814 (respectivamente).

QUADRA "20", com a área de 11.301,01m², constituída por 30 (trinta) lotes, numerados de 01 a 30, compreendidos entre as Ruas: Rua Rio Aporé, Rua Rio Paraná, Rua Rio Brilhante, Rua Rio Pardo, sendo os Lotes n. 01 e 02, com frente para a Rua Rio Aporé, possuindo o Lote n. 01 área de 360,00m², Lote n. 02, área de 440,32m², os Lotes n. 03 a 06, com frente para a Rua Rio Pardo, possuindo os Lotes n. 03 a 04 área de 414,60m², cada lote, Lote n. 05, área de 409,23m², os Lotes n. 06 e 15, com frente para a Rua Rio Brilhante, possuindo os Lotes n. 06 a 15, área de 360,00m², cada lote; os Lotes n. 16 a 20, com frente para a Rua Rio Paraná, possuindo o Lote n. 16 a área de 409,23m², Lotes n. 17 a 19 a área de 414,60m², cada lote, Lote n. 20, área de 409,23m²; e os Lotes n. 21 a 30, com frente para a Rua Rio Aporé, possuindo os Lotes n. 21 a 30 a área de 360,00m², cada lote; conforme matrículas n. 72.815 a 72.844 (respectivamente).

QUADRA "21", com a área de 11.107,32m², constituída por 30 (trinta) lotes, numerados de 01 a 30, compreendidos entre as Ruas: Rua Rio Aporé, Rua Rio Negro, Rua Rio Brilhante, Rua Rio Paraná, sendo os Lotes n. 01 a 05, com frente para a Rua Rio Paraná, possuindo o Lote n. 01 área de 387,51m², Lotes n. 02 a 04, área de 392,88m², cada lote, Lote n. 05, área de 387,51m²; os Lotes n. 06 a 15, com frente para a Rua Rio Brilhante, possuindo os Lotes n. 06 a 15, área de 360,00m², cada lote; os Lotes n. 16 a 20, com frente para a Rua Rio Negro, possuindo o Lote n. 16 a área de 387,51m², Lotes n. 17 a 19 a área de 392,88m², cada lote, Lote n. 20, área de 387,51m²; e os Lotes n. 21 a 30, com frente para a Rua Rio Aporé, possuindo os Lotes n. 21 a 30 a área de 360,00m², cada lote; conforme matrículas n. 72.845 a 72.874 (respectivamente).

QUADRA "22", com a área de 7.213,37m², constituída por 17 (dezessete) lotes, numerados de 01 a 17, compreendidos entre as Ruas: Avenida Rio Miranda, Rua Rio Pardo, Rua Rio Ivinhema, sendo os Lotes n. 01 a 06, com frente para a Rua Rio Ivinhema, possuindo o Lote n. 01 área de 713,33m²; Lotes n. 02 a 05, área de 360,00m², cada lote, Lote n. 06, área de 444,00m²; os Lotes n. 07 a 13, com frente para a Rua Rio Pardo, possuindo o Lote n. 07 área de 384,63m², Lotes n. 08 a 12, área de 360,00m², cada lote; Lote n. 13, área de 716,52m²; e os Lotes n. 14 a 17, com frente para a Avenida Rio Miranda, possuindo o Lote n. 14 a área de 516,78m², Lote n. 15 a área de 479,70m², Lote n. 16, área de 356,64m², Lote n. 17, área de 361,77m²; conforme matrículas n. 72.875 a 72.891 (respectivamente).

QUADRA "23", com a área de 11.324,52m², constituída por 30 (trinta) lotes, numerados de 01 a 30, compreendidos entre as Ruas: Rua Rio Brilhante, Rua Rio Paraná, Rua Rio Ivinhema, Rua Rio Pardo, sendo os Lotes n. 01 a 05, com frente para a Rua Rio Pardo, possuindo o Lote n. 01 área de 409,23m², Lotes n. 02 a 04, área de 414,60m², cada lote, Lote n. 05, área de 409,23m²; os Lotes n. 06 a 15, com frente para a Rua Rio Ivinhema, possuindo os Lotes n. 06 a 15, área de 360,00m², cada lote; os Lotes n. 16 a 20, com frente para a Rua Rio Paraná, possuindo o Lote n. 16 a área de 409,23m², Lotes n. 17 a 19 a área de 414,60m², cada lote, Lote n. 20, área de 409,23m²; e os Lotes n. 21 a 30, com frente para a Rua Rio Brilhante, possuindo os Lotes n. 21 a 30 a área de 360,00m², cada lote; conforme matrículas n. 72.892 a 72.921 (respectivamente).

QUADRA "24", com a área de 11.107,32m², constituída por 30 (trinta) lotes, numerados de 01 a 30, compreendidos entre as Ruas: Rua Rio Brilhante, Rua Rio Negro, Rua Rio Ivinhema, Rua Rio Paraná, sendo os Lotes n. 01 a 05, com frente para a Rua Rio Paraná, possuindo o Lote n. 01 área de 387,51m², Lotes n. 02 a 04, área de 392,88m², cada lote, Lote n. 05, área de 387,51m²; os Lotes n. 06 a 15, com frente para a Rua Rio Ivinhema, possuindo os Lotes n. 06 a 15, área de 360,00m², cada lote; os Lotes n. 16 a 20, com frente para a Rua Rio Negro, possuindo o Lote n. 16 a área de 387,51m², Lotes n. 17 a 19 a área de 392,88m², cada lote, possuindo o Lote n. 16 a área de 387,51m²; e os Lotes n. 21 a 30, com frente para a Rua Rio Brilhante, Lote n. 20, área de 387,51m², possuindo os **Lotes n. 21 a 30** a área de 360,00m³, cada lote; **conforme matrículas n. 72.922 a 72.951 (respectivamente)**.

QUADRA "25", com a área de 11.188,42m³, constituída por 29 (vinte e nove) lotes, numerados de 01 a 29, compreendidos entre as Ruas: Rua Rio Ivinhema, Rua Rio Pardo, Rua Rio Paraguai, Rua Quixeramobim, sendo os **Lotes n. 01 a 04**, com frente para a Rua Quixeramobim, possuindo o **Lote n. 01** área de 687,73m³, **Lotes n. 02 e 03**, área de 414,60m², cada lote, **Lote n. 04**, área de 409,23m²; os **Lotes n. 05 a 14**, com frente para a Rua Rio Paraguai, possuindo os **Lotes n. 05 a 14**, área de 360,00m², cada lote; os **Lotes n. 15 a 18**, com frente para a Rua Rio Pardo, possuindo o **Lote n. 15** a área de 409,23m², **Lotes n. 16 a 18** a área de 414,60m³, cada lote, **Lote n. 19**, área de 409,23m²; e os **Lotes n. 20 a 29**, com frente para a Rua Rio Ivinhema, possuindo os **Lotes n. 20 a 29** a área de 360,00m², cada lote; **conforme matrículas n. 72.952 a 72.980 (respectivamente)**.

QUADRA "26", com a área de 11.324,52m³, constituída por 30 (trinta) lotes, numerados de 01 a 30, compreendidos entre as Ruas: Rua Rio Ivinhema, Rua Rio Paraguai, Rua Rio Paraná, Rua Rio Pardo, sendo os **Lotes n. 01 a 05**, com frente para a Rua Rio Pardo, possuindo o **Lote n. 01** área de 409,23m², **Lotes n. 02 a 04**, área de 414,60m², cada lote, **Lote n. 05**, área de 409,23m²; os **Lotes n. 06 a 15**, com frente para a Rua Rio Paraguai, possuindo os **Lotes n. 06 a 15**, área de 360,00m², cada lote; os **Lotes n. 16 a 20**, com frente para a Rua Rio Paraná, possuindo o **Lote n. 16** a área de 409,23m², **Lotes n. 17 a 19** a área de 414,60m², cada lote, **Lote n. 20**, área de 409,23m²; e os **Lotes n. 21 a 30**, com frente para a Rua Rio Ivinhema, possuindo os **Lotes n. 21 a 30** a área de 360,00m², cada lote; **conforme matrículas n. 72.981 a 73.010 (respectivamente)**.

QUADRA "27", com a área de 11.107,32m³, constituída por 30 (trinta) lotes, numerados de 01 a 30, compreendidos entre as Ruas: Rua Rio Paraná, Rua Rio Ivinhema, Rua Rio Negro, Rua Rio Paraguai, sendo os **Lotes n. 01 a 05**, com frente para a Rua Rio Paraná, possuindo o **Lote n. 01** área de 387,51m³, **Lotes n. 02 a 04**, área de 392,88m³, cada lote, **Lote n. 05**, área de 387,51m³; os **Lotes n. 06 a 15**, com frente para a Rua Rio Paraguai, possuindo os **Lotes n. 06 a 15**, área de 360,00m³, cada lote; os **Lotes n. 16 a 20**, com frente para a Rua Rio Negro, possuindo o **Lote n. 16** a área de 387,51m³, **Lotes n. 17 a 19** a área de 392,88m³, cada lote, **Lote n. 20**, área de 387,51m³; e os **Lotes n. 21 a 30**, com frente para a Rua Rio Ivinhema, possuindo os **Lotes n. 21 a 30** a área de 360,00m², cada lote; **conforme matrículas n. 73.011 a 73.040 (respectivamente)**.

QUADRA "28", com a área de 11.324,52m³, constituída por 30 (trinta) lotes, numerados de 01 a 30, compreendidos entre as Ruas: Rua Quixeramobim, Rua Alecrim, Rua Rio Pardo, Rua Rio Paraguai, sendo os **Lotes n. 01 a 05**, com frente para a Rua Quixeramobim, possuindo o **Lote n. 01** área de 409,23m³, **Lotes n. 02 a 04**, área de 414,60m², cada lote, **Lote n. 05**, área de 409,23m²; os **Lotes n. 06 a 15**, com frente para a Rua Alecrim, possuindo os **Lotes n. 06 a 15**, área de 360,00m², cada lote; os **Lotes n. 16 a 20**, com frente para a Rua Rio Pardo, possuindo o **Lote n. 16** a área de 409,23m², **Lotes n. 17 a 19** a área de 414,60m², cada lote, **Lote n. 20**, área de 409,23m²; e os **Lotes n. 21 a 30**, com frente para a Rua Rio Paraguai, possuindo os **Lotes n. 21 a 30** a área de 360,00m³, cada lote; **conforme matrículas n. 73.041 a 73.070 (respectivamente)**.

QUADRA "29", com a área de 11.324,52m³, constituída por 30 (trinta) lotes, numerados de 01 a 30, compreendidos entre as Ruas: Rua Rio Pardo, Rua Alecrim, Rua Rio Paraná, Rua Rio Paraguai, sendo os **Lotes n. 01 a 05**, com frente para a Rua Rio Pardo, possuindo o **Lote n. 01** área de 409,23m³, **Lotes n. 02 a 04**, área de 414,60m², cada lote, **Lote n. 05**, área de 409,23m²; os **Lotes n. 06 a 15**, com frente para a Rua Alecrim, possuindo os **Lotes n. 06 a 15**, área de 360,00m², cada lote; os **Lotes n. 16 a 20**, com frente para a Rua Rio Paraná, possuindo o **Lote n. 16** a área de 409,23m², **Lotes n. 17 a 19** a área de 414,60m³, cada lote,

178

Lote n. 20, área de 409,23m²; e os **Lotes n. 21 a 30**, com frente para a Rua Rio Paragual, possuindo os **Lotes n. 21 a 30** a área de 360,00m², cada lote; <u>conforme matrículas n. 73.071 a 73.100 (respectivamente)</u>.

QUADRA "30", com a área de 11.107,32m², constituída por 30 (trinta) lotes, numerados de 01 a 30, compreendidos entre as Ruas: Rua Rio Paraná, Rua Alecrim, Rua Rio Negro, Rua Rio Paragual, sendo os **Lotes n. 01 a 05**, com frente para a Rua Rio Paraná, possuindo o Lote n. 01 área de 387,51m², **Lotes n. 02 a 04**, área de 392,88m², cada lote, **Lote n. 05**, área de 387,51m²; os **Lotes n. 06 a 15**, com frente para a Rua Alecrim, possuindo os **Lotes n. 06 a 15**, área de 360,00m², cada lote; os **Lotes n. 16 a 20**, com frente para a Rua Rio Negro, possuindo o **Lote n. 16** a área de 387,51m², **Lotes n. 17 a 19** a área de 392,88m², cada lote, Lote n. 20, área de 387,51m²; e os **Lotes n. 21 a 30**, com frente para a Rua Rio Paragual, possuindo os **Lotes n. 21 a 30** a área de 360,00m², cada lote; <u>conforme matrículas n. 73.101 a 73.130 (respectivamente)</u>.

QUADRA "31", com a área de 2.179,07, constituída por 02 (dois) lotes, compreendidos entre as Ruas: Rua Quixeramobim, Avenida Rio Miranda, sendo o **Lote n. 01**, com frente para a Rua Quixeramobim, possuindo o **Lote n. 01** área de 394,50m², e **ÁREA VERDE 01**, com frente para a Avenida Rio Miranda, com a área de 1.784,57m², <u>conforme matrículas n. 73.131 e 73.132 (respectivamente)</u>.

ÁREA VERDE 02", com a área de 549,90m², compreendida entre a divisa do lote 01 da quadra 12 – Área Verde, matrícula 59.582, livro 02, deste Registro Imobiliário, e Avenida Rio Miranda, <u>conforme matrícula n. 73.133</u>.

A área institucional de 8,00%, equivalente a 38.720,00m², será doada na matrícula n. 63.024 do livro 02, deste Registro Imobiliário, conforme TAC com o Ministério Público.

As áreas Verdes passaram a integrar ao domínio do Município de Três Lagoas/MS, conforme Artigo 22, da Lei 6.766 de 19/12/1979. Emolumentos R$ 12.618,00; FUNJECC 10% R$ 1.261,80; FUNJECC 3% R$ 378,54; FUNADEP 6% R$ 757,08; FUNDE-PGE 4% R$ 504,72. Selo digital n. AIV63155-018 (este selo poderá ser conferido e autenticado no site: www.tjms.jus.br/corregedoria/selos/pesquisaoSelo.php). Eu, Fernanda Andrade Moura de Almeida, auxiliar extrajudicial, digitei. Eu, Giovani Gomes Roman, escrevente autorizado, conferi. Dou fé. Três Lagoas/MS, 19 de fevereiro de 2015. Oficial/Substituto/Escrevente Autorizado. –

Av.04/M.62.714.- "Ex-Officio".- Faz-se a presente averbação nos termos do art. 213, inc.I, alínea "e" da Lei 6.015/73, para constar que, por um lapso, quando do registro n. 03 desta matrícula constou erroneamente que a área do lote 05 da quadra 11 é de 377,02m², quando na realidade o certo é 373,02m², conforme faz prova mapa do loteamento arquivado neste RI. Eu, Bruno Sperigone da Silva, auxiliar extrajudicial, digitei e conferi. Dou fé. Três Lagoas/MS, 15 de maio de 2015. Oficial/Substituto/Escrevente Autorizado. –

CERTIFICO que a presente fotocópia confere com a matrícula original de **n. 62.714** e que, nos termos do disposto artigo 19, § 1º da Lei 6.015/1973, tem valor de certidão. O referido é verdade e dou fé. **Três Lagoas, MS, em 14 de dezembro de 2021.** SELO nº AFU68118-695-NOR.

____
Oficial do Registro / Substituto / Escrevente

Acesse o site http://www.tjms.jus.br/corregedoria/selos/pesquisaselos.php para visualizar a autenticidade desta certidão.

Pedido de certidão nº: 248224

Controle: Página: 0010/0010

ANNEX 3 - REGISTRATION 46.928

REPÚBLICA FEDERATIVA DO BRASIL
ESTADO DE MATO GROSSO DO SUL
SERVIÇO DE REGISTRO DE IMÓVEIS

AV. ANTONIO TRAJANO, 439 - CEP 79600-005 - FONE nº 3521-2247 - FAX nº 3521-9247 TRÊS LAGOAS-MS

IMÓVEL:- Rural, denominado "Fazenda Santa Luzia - Gleba "A", com a área de 10,2028 has (dez hectares, vinte ares e vinte e oito centiares), Perímetro: 1.454,27m, situado neste município e comarca de Três Lagoas-MS., dentro das seguintes medidas e confrontações: "Inicia-se a descrição deste perímetro no vértice AAC-M-N427, de coordenadas E 429.995,223m e N 7.699.313,156m, situado na divisa do perímetro urbano de Três Lagoas/MS com a Fazenda Santa Luzia (Parte) – Gleba "C", da Agropecuária Prata Tibery Ltda., matrícula: 12.364; deste, segue confrontando com a Fazenda Santa Luzia (Parte) - Gleba "C" com o seguinte azimute e distância: 120°25'33" e 486,59m até o vértice AAC-M-N426, de coordenadas E 430.414,799m e N 7.699.066,739m, situado na divisa da Fazenda Santa Luzia (Parte) - Gleba "C" e junto ao limite da faixa de domínio da Estrada de Ferro Novoeste Ferropasa – Ferronorte Participações S/A, que liga Três Lagoas/MS a Campo Grande/MS; deste, segue junto ao limite da faixa de domínio da referida estrada de ferro com os seguintes azimutes e distâncias: 236°16'55" e 125,91m até o vértice AAC-M-C498, de coordenadas E 430.310,068m e N 7.698.996,044m; 227°45'25" e 109,17m até o vértice AAC-M-C499, de coordenadas E 430.229,252m e N 7.698.923,454m; 219°18'25" e 167,35m até o vértice AAC-M-C490, de coordenadas E 430.123,238m e N 7.698.793,961m, situado junto ao limite da faixa de domínio da Estrada de Ferro Novoeste Ferropasa – Ferronorte Participações S/A e na divisa da Chácara Paulista, de José Teixeira de Oliveira, matrícula: 10.963; deste, segue confrontando com a Chácara Paulista com o seguinte azimute e distância: 339°59'45" e 484,29m até o vértice AAC-M-C491, de coordenadas E 429.957,566m e N 7.699.249,035m, situado na divisa da Chácara Paulista com o perímetro urbano de Três Lagoas/MS; deste, segue confrontando com o perímetro urbano de Três Lagoas/MS com os seguintes azimutes e distâncias: 340°01'20" e 15,27m até o vértice AAC-M-C492, de coordenadas E 429.952,350m e N 7.699.263,386m; 40°44'34" e 85,69m até o vértice AAC-M-N427, ponto inicial da descrição deste perímetro. Todas as coordenadas aqui descritas estão georreferenciadas ao Sistema Geodésico Brasileiro a partir do vértice Geodésico da Rede GPS do Mato Grosso do Sul "MS22" de coordenadas N 7.704.014,261m e E 428.017,158m (situado no fuso 51° W.Gr") e RBMC (Rede Brasileira de Monitoramento Contínuo): vértice geodésico de Cuiabá "CUIB-92583", de coordenadas N 8.280.082,107m e E 599.791,609m (situado no fuso 57° W.Gr"), vértice geodésico de Brasília "BRAZ-91200" de coordenadas N 8.234.791,575m e E 191.946,760m (situado no fuso 45° W.Gr") e do vértice geodésico de Presidente Prudente "UEPP-91559" de Coordenadas N 7.553.888,233m e E 457.915,946m (situado no fuso 51° W.Gr"), sendo que as coordenadas desse imóvel encontram-se representadas no Sistema UTM, referenciadas ao Meridiano Central 51° W.Gr", tendo como datum o SAD-69. Todos os azimutes e distâncias, área e perímetro foram calculadas no plano de projeção UTM. Memorial descritivo datado de 24/07/2007, elaborado e assinado pelo Engenheiro Cartógrafo: Mário Maurício Vasquez Beltrão – CREA 1.577/D-MS, credenciado no INCRA sob código: AAC, ART nº 001472 E. Apresentou Certidão Negativa de Débitos de Imóvel Rural, emitida em 26/07/2007, pela Secretaria da Receita Federal, com validade até 28/01/2008, código de controle da certidão: 39CA.EA0E.0857.1838, onde consta o NIRF 0.381.994-9, Fazenda Santa Luzia, área total 631,5 ha, município de Três Lagoas-MS, em nome de Wania Ruth Prata Tibery. CCIR- 2003/2004/2005, onde consta o código do imóvel 9120340191607, área total 1.181,6000, mód rural. 40,0086, nº mód.rurais: 23,10 mód.fiscal 35,0000, nº mód.fiscais 33,76, FMP 2,0000, classificação fundiária: grande propriedade produtiva. Fazenda Santa Luzia, área registrada 1.181,5000 has, município de Três Lagoas-MS, em nome de Ascêncio Garcia Lopes, brasileiro, código da pessoa: 004067053. Registro Anterior:- Matrícula 46.927, livro 02, deste Registro Imobiliário. **PROPRIETÁRIA:**- **AGROPECUÁRIA PRATA TIBERY LTDA** empresa jurídica de direito privado, inscrita no CNPJ/MF nº 06.713.380/0001-96, com endereço comercial na Avenida Ademar Pereira de Barros, nº 872, Jardim Bela Suíça, na cidade de Londrina-PR. Emol. R$ 16,00; FUNJECC 10% que incide sobre o emolumento, correspondente a R$ 1,60, recolhido pelo usuário; FUNJECC 3% que incide sobre o emolumento, correspondente a R$ 0,48, recolhido pela serventia. Dou fé. Três Lagoas-MS, 23 de agosto de 2007. Aimée Aparecida de Souza Ferreira, Interventora.-

Av.01/M.46.928.-De acordo com a averbação nº 05 da matrícula 12.364, e averbação nº 01 da matrícula 46.927, ambas livro 02, deste registro imobiliário, existe no imóvel objeto da presente

"Feliz a Nação cujo Deus é o Senhor" —Continua no verso...

matrícula 20% (Vinte por cento) de Reserva Legal, em cumprimento ao que determina a Lei 4.771 de 15/09/65, com as alterações da Lei 7.803, de 18/07/89, não podendo dela ser feita qualquer tipo de exploração, a não ser mediante autorização do IBAMA. Dou fé. Três Lagoas-MS.,23 de agosto de 2007. Aimée Aparecida de Souza Ferreira, Interventora.-

R.02/M.46.928.-Protocolo:-131.609 em 10/12/2007.-Venda e Compra.-Pela Escritura Pública de Venda e Compra, lavrada no livro 69-E, fls. 022/23vº, em 30 de novembro de 2007, pelo Cartório Distrital de Tamarana, comarca de Londrina-PR, a proprietária AGROPECUÁRIA PRATA TIBERY LTDA, já qualificada, vendeu do imóvel objeto da presente matrícula, somente uma parte ideal, correspondente à 0,50% (meio por cento) à PLAENGE CONSTRUÇÕES LTDA, pessoa jurídica de direito privado, inscrita no CNPJ/MF sob nº 08.734.878/0001-62, com sede na Avenida Tiradentes, nº 1.000, sala 05, na cidade de Londrina-PR, e 0,50% (meio por cento) à VECTRA CONSTRUTORA LTDA, pessoa jurídica de direito privado, inscrita no CNPJ/MF sob nº 01.263.107/0001-95, com sede na Avenida Inglaterra, nº 385, Sala 35, na cidade de Londrina-PR, pelo preço de R$ 5.579,72 (CINCO MIL, QUINHENTOS E SETENTA E NOVE REAIS E SETENTA E DOIS CENTAVOS). Apresentou pagamento do ITBI no valor de R$ 111,59, sobre 2% de R$ 5.579,72, conforme DAM nº 49787468, expedida em 10 de dezembro de 2007, pela Prefeitura Municipal local. Certidão Negativa de Débitos Relativos ao Imposto sobre a Propriedade Territorial Rural, emitida em 31/12/2007, pela Secretaria da Receita Federal do Brasil, com validade até 30/06/2008, código de controle da certidão: 54F2.2F38.E353.8435, onde consta o NIRF 7.390.367-1, Fazenda Santa Luzia – Gleba A, área total 10,2 has, município de Três Lagoas-MS., em nome de Agropecuária Prata Tibery Ltda. Certidão Negativa de Débito nº 636124, emitida em 09/01/2008 pelo IBAMA, com validade até 08/02/2008, em nome de Agropecuária Prata Tibery Ltda. CCIR-2003/2004/2005, o mesmo constante da abertura desta matrícula. Emol. R$ 100,00; FUNJECC 10% que incide sobre o emolumento, correspondente a R$ 10,00, recolhido pelo usuário; FUNJECC 3% que incide sobre o emolumento, correspondente a R$ 3,00, recolhido pela serventia. Selo de Autenticidade nº ACT 06497. Dou fé. Três Lagoas-MS.,10 de janeiro de 2008. Aimée Aparecida de Souza Ferreira, Interventora.-

Av.03/M.46.928.-Protocolo:-149.695 em 21/10/2010.-Descaracterização do Imóvel.- Pelo requerimento datado de 04 de outubro de 2010, os proprietários AGROPECUÁRIA PRATA TIBERY LTDA; PLAENGE CONSTRUÇÕES LTDA e VECTRA CONSTRUTORA LTDA, requerem a presente averbação para constar que de acordo com Declaração expedida em 06/10/2010, pela Prefeitura Municipal de Três Lagoas/MS, assinada pelo diretor de Departamento de Tributação, Divino Teodoro dos Santos e pelo Secretário Municipal de Finanças, Planejamento e Construção Geral, Walmir Marques Arantes, onde declara que o imóvel objeto dessa matrícula encontra-se no perímetro urbano, bem como Declaração expedida em 14 de outubro de 2010, devidamente assinada pelo Chefe Substituto de Divisão de Ord. Da Estrutura Fundiária do Instituto Nacional de Colonização e Reforma Agrária – INCRA – Superintendência Regional de Mato Grosso do Sul – SR (16) MS – Divisão de Ordenamento da Estrutura Fundiária, Sr. Jose Carlos V. Antunes – Port./INCRA/D.A/n. 62/08, foi cancelado o código do INCRA sob n. 950.114.274.143-3, do imóvel objeto da presente matrícula ficando portanto descaracterizado de imóvel rural para imóvel URBANO. Emol. R$ 34,00; FUNJECC 10% R$ 3,40; FUNJECC 3% R$ 1,02. Selo de autenticidade n. ADN 38549. Eu, Douglas Rodrigo Damasceno Fernandes, Auxiliar Extrajudicial, digitei e conferi. Dou fé. Três Lagoas/MS, 11 de novembro de 2010. Oficial/Substituto/Escrevente Autorizado.

-CONTINUA NA FOLHA 02-.

R.04/M.46.928.- Protocolo:- 155.470 em 15/07/2011. — **Compra e venda**. - Pela escritura pública de compra e venda lavrada no livro 80-E, fle. 005/006 v, em 07 de julho de 2011, pelo Cartório Distrital da cidade de Tamarana, comarca de Londrina/PR, a proprietária **PLAENGE CONSTRUÇÕES LTDA**, já qualificada, representada por seus administradores, Alexandre Dores Fabian, RG 1.684.131-SSP/PR, CPF/MF 542.300.479-91, brasileiro, casado, engenheiro civil, com endereço profissional na Av. Tiradentes, n. 1.000, sala 02, na cidade de Londrina/PR; Carlos Roberto da Silva Melquiades, RG 748.287-SSP/PR, CPF/MF 313.019.169-00, brasileiro, casado, economista, residente e domiciliado na Av. Harry Prochet, n. 677, casa 06, na cidade de Londrina/PR; **vendeu a parte ideal correspondente a 0,50% (meio por cento)** do imóvel objeto da presente matrícula a **VECTRA CONSTRUTORA LTDA**, CNPJ/MF 01.263.107/0001-95, pessoa jurídica de direito privado, com sede e foro na Avenida Inglaterra, n. 385, sala 35, na cidade de Londrina/PR, representada por, Roberta Costa Alves Nunes Mansano, RG 4.899.778-3-SSP/SP, CPF/MF 030.649.429-99, brasileira, casada, empresaria, residente e domiciliada na Av. Adhemar Pereira Barros, n. 1.200, apt. 1.902, na cidade de Londrina/PR; pelo preço de R$ 3.300,00 (TRÊS MIL E TRENTOS REAIS), comparecendo como anuente: Agropecuária Prata Tibery Ltda., já qualificada, ato representada por seu procurador Luiz Henrique Prata Tibery Garcia Lopes, RG 1.336.374-9-SSP/PR, CPF/MF 533.570.869-72, brasileiro, casado, engenheiro agrônomo, residente e domiciliado na Rua Santiago, n. 737, na cidade de Londrina/PR. Apresentou o pagamento do ITBI conforme guia n. 78560125, expedida em 13/07/2011, pela Prefeitura Municipal local. Certidão Negativa de Debito protocolada sob n. 62948, e expedida em 13 de julho de 2011, pela referida Prefeitura. (Valor venal da parte ideal de 0,5%, no total de R$ 3.300,00, avaliado pela Prefeitura Municipal local). Imóvel cadastrado na Prefeitura sob n. 7.73.000.0012.00001. Emolumentos: R$ 56,00; FUNJECC 10% R$ 5,60; FUNJECC 3% R$ 1,68. Selo digital n. ABE 32409-696. Eu, Douglas Rodrigo Damasceno Fernandes, auxiliar extrajudicial, digitei e conferi. Dou fé. Três Lagoas/MS, 11 de agosto de 2011. Oficial/Substituto/Escrevente autorizado.- _______________

R.05/M.46.928.- Protocolo:- 160.466 em 15/02/2012. -**LOTEAMENTO FECHADO TERRAS DE JUPIÁ**. - O imóvel objeto desta matrícula foi totalmente LOTEADO, conforme planta e memoriais descritivos aprovados pela Secretaria de Obras da Prefeitura Municipal local, de acordo com o Ato de Aprovação de Loteamento n. 009/2.011 de 15 de dezembro de 2011, e Alvará de Licença n. 047/2.012 de 01 de fevereiro de 2012, assinado pelo Secretario de Obra Engº Rodrigo Pelho Rizzo- CREA VISTO MS — 14.095, conforme consta do processo arquivado neste cartório. O Loteamento, de propriedade de **AGROPECUÁRIA PRATA TIBERY LTDA** e **VECTRA CONSTRUTORA LTDA**, com a denominação de ("TERRAS DE JUPIÁ") contém a área total de 102.028,00m², assim distribuídos:

I - ÁREA LOTEADA E VENDÁVEL: 52.200,12 m²., subdividida em (07) quadras, assim designadas:- **QUADRA – 01**, com a área de 13.682,62 m²., constituída por (44) lotes, numerados de 01 a 44; matrícula 60.589 a 60.632; **QUADRA – 02**, com a área de 11.794,73 m²., constituída por (38) lotes, numerados de 01 a 38, matrícula 60.633 a 60.670; **QUADRA – 03**, com a área de 2.803,88 m²., constituída por (09) lotes, numerados de 01 a 09, matrícula 60.671 a 60.679; **QUADRA – 04**, com a área de 8.263,92 m²., constituída por (26) lotes, numerados de 01 a 26, matrícula 60.680 a 60.705; **QUADRA – 05**, com a área de 8.100,00 m²., constituída por (27) lotes, numerados de 01 a 27, matrícula 60.706 a 60.732; **QUADRA – 06**, com a área de 5.111,64 m²., constituída por (17) lotes, numerados de 01 a 17, matrícula 60.733 a 60.749; **QUADRA – 08**, com a área de 2.443,33 m²., constituída por (07) lotes, numerados de 01 a 07; matrícula 60.750 a 60.756;

II - ÁREA DE USO COMUM:- 5.106,22 m²., destinados a área de uso comum, denominadas **L.U.C. 01**, com a área de 579,72 m²., matrícula 60.757; **L.U.C. 02**, com a área de 319,86 m²., matrícula. 60.758; **L.U.C. 03**, com a área de 324,63 m²., matrícula 60.759; **L.U.C. 04**, com a

183

REPÚBLICA FEDERATIVA DO BRASIL
ESTADO DE MATO GROSSO DO SUL
SERVIÇO DE REGISTRO DE IMÓVEIS

Miriam Reis Costa
Oficial de Registro de Imóveis

Eduard Reis Costa Filho   *Miriam Cilene Reis Costa*   *Miriam Clarice Reis Costa*   *Giovani Gomes Rponeu*   *Sabrina de Lima Valim*
Oficial Substituto Designado   2º Oficial Substituto   3º Oficial Substituto   Escrevente Habilitada   Escrevente de Registro

Av. Antonio Trajano, 1320, TRÊS LAGOAS/ MS – Fone/ Fax: (67) 3521224

CERTIFICO, em razão de meu cargo e por requerimento de parte interessada, que, revendo neste Cartório o fichário e **Livro 3-A, f. 112**, verifiquei constar o registro com o seguinte teor:

N. de Ordem: 462 Data: 22 de junho de 1922. **Circunscrição:** Município de Três Lagoas. **Denominação ou rua e Número do Imóvel:** Três Lagoas. **Característicos e Confrontações:** Lote de terras com a superfície de 3.659 hectares, reservado para patrimônio da povoação de Três Lagoas pelo Decreto da Presidência do Estado n. 311 de 9 de abril de 1912, confinando ao Norte com propriedades da companhia Feira de Gado de Três Lagoas e com a Fazenda Varginha pertencente a herança de Delfino Pereira dos Santos; ao Poente com a fazenda Córrego do Pinto, pertencente a diversos; ao Sul com o leito da Estrada de Ferro Noroeste do Brasil e ao Nascente com a referida propriedade da Companhia Feira de Gado de Três Lagoas estando o perímetro demarcado com (50) cinquenta marcos dispostos pela forma declarada no título. Eu, Odorico Alves Corrêa, sub-oficial o escrevi. O Oficial do Registro. (ass) Felipe Nery Monteiro. **Nome, Domicílio, Profissão, Estado e Residência do Adquirente: O MUNICÍPIO DE TRÊS LAGOAS**, do Estado de Mato Grosso, com sede na cidade de Três Lagoas. **Nome, Domicílio, Profissão, Estado e Residência do Transmitente: O ESTADO DE MATO GROSSO** com capital na cidade de Cuiabá. **Título de transmissão:** Concessão. **Forma do título, Data e Serventuário:** Título definitivo expedido em 13 de março de 1922 pela repartição de Terras, Minas, colonização do estado, subscrito pelo Diretor da Repartição agrimensor Otavio de Vasconcelos Neves, assinado pelo Presidente do Est. Coronel Pedro Celestino Corrêa, e pelo Secretário da Agricultura Dr. Carlos Borralho. **Valor do Contrato:** Não tem. **Condições do Contrato:** Servir o lote concedido de patrimônio a povoação de Três Lagoas, que se formava ao tempo da concessão. **Averbações:** Certifico que por Certidão expedida pela Secretaria da Prefeitura Municipal desta Cidade, em data de 18 de agosto de 1977, o imóvel objeto do registro em frente, corresponde atualmente ao loteamento e planejamento desta cidade e seu respectivo zoneamento, tudo de acordo com dita certidão. O Referido é verdade e dou fé. Eu, Edith Martins Lopes escrevente habilitada, escrevi, conferi, subscrevo e assino. Três Lagoas, 26 de agosto de 1977. (ass) Edith Martins Lopes. Aforados 1.000m², do lote 3, quart. 52 da 2ª Z. urbana, a Jadanor de Oliveira Santos, conf. reg. 1- 3.364, fls. 72, 2-Q. Aforados os lotes n. 3, 4, 5 e 8 da q. 10 da 3ª zona urbara C/ 4.000m², conf. reg. 1-3.577, fls. 85, L. 2-R. Aforados 1.000m² do lote 9, q.57, 1ª zona urbana a José Francisco, conf. reg. n. 1- 3.578, fls. 86 L. 2-R. Aforado o lote n. 2, q. 44, 4ª z. urbana á José B. dos Santos conf. reg. n. 1-3.878, fls. 190, L. 2-S, de 13/12/77. Aforado o lote n. 6, q. 110, 3ª zona urb. a Melquides Ferreira Neves, conf. reg. n. 1-4.162, fls. 75, L. 2-U, 13/2/78. Aforado o lote n. 8, q. 97, 3ª z. urb. a José Amâncio Veron, conf. reg. 1-4.681, L. 2-X, fls. 197, 18/05/78. Aforado lotes 4 e 5 da q. 6, 3ª z. urb. desta cidade, c/ 2.000ms², conf. reg. 1-4.738, L.2-Z, fls.56. Aforado lotes de 1 a 5 q. 92, conf. reg. 1-5.132 fls. 45, L. 2-AB a Davina Alves de Souza. Aforado o lote 17, q. 131, 1ª z. urb. conf. Reg.1-5.155, fls. 69, 2-AB, a Belmiro P. Ramos. Aforado o lote n. 45 c/ 48.500m², 2ª z. suburbana conf. reg. n. 1- 5.168, fls. 82, 2-AB. Aforado lote 7, q. 44/45, c/ 419,37m conf. reg. 1-5.342, fls. 57, 2-AC em 08/9/78. Aforado o lote suburb. 112 c/ 48.400m³ conf. reg. 1- 5.437, fls. 154 2-AC em 27/09/78. Aforado o lote 10, q. 110 c/ 1.000m², conf. reg. 1-5.402, L.

Em tudo dai graças
I Ts S.18

2-AC, fls. 119 em 20-09-78. Aforado o lote 10. q. 02 c/ 800,00m², conf. reg. 1- 5.672, L. 2- AD, fls. 191, em 13/11/78. Aforados lotes 1 e 2, q. 99, 3ª z. urbana, c/ 2.000ms, Reg. 1-5.783, fls. 103, L. 2-AE. Aforado lote 3 e 4, q. 78, 4ª z. urbana, c/ 2000m². reg. 1- 5.875, fls. 199, L. 2-AE. Aforado lote 08, q. 98, 3ª z. urbana, c/ 1.000m² Reg. 1-5.957, fls. 81, L. 2-AF. Aforado lote 08. q. 75, 1ª z. urbana c/ 250,00m² R. 1- 8.029, fls. 157, L. 2-AF. Aforado lote 8 q. 135. 4ª z. urbana c/ 1.000m² R. 1- 6.308, fls. 43, L. 2-AH. Aforado lote 5 q. 44, 4ª z. urbana c/ 500m2 R. 1-6.499, fls. 37, L. 2-AI. Aforado lote 2 q. 48. 1ª z. urbana c/ 500m² R. 1-7.282, fls. 48, L. 2-AH. Aforado lote 5. q. 41, 4ª z. urbana c/ 500m² R. 1-7.598, fls. 158 L. 2-AO. Aforado lote 2, q. 44, 4ª z. urbana c/ 1.000m² R. 1- 7.757, fls. 119, 2-AP. Aforado lote 10 q. 134, 1ª z. urbana c/ 500m² R. 1-7.983 fls. 150, L. 2-AQ. Aforado lote 130 suburbano c/ 13.855m² R. 1- 8.058, fls. 25 L. 2-AR. Aforado lote 4 e 5 q. 90, 3ª z. urbana c/ 2.000m², R. 1-7.698, fls. 59 L. 2-AP. Aforado lote 1, 2, 3 e 4 q. 04 2ª z. urbana c/ 1.800m² R.1-9.774, fls. 169 L. 2-BB. Aforado lote 6 q. 4-A 2ª z. urbana c/ 1.100m² R. 1-9.775. fls. 170. L. 2- BB. Aforado lote 03, q. 01, 2ª z. urbana c/ 275m2 R. 1-9.940, fls. 135 2-BC. Aforado lote 2 q. 3, 2ª z. urbana c/ 650m² R. 1-9960 fls. 155 2-BC. Aforado lote 14 q. 76, 1ª z. urbana c/ 17x50m, R. 1-10.253, fls. 54 L. 2-BE. Aforado lote 3 q. 43 4ª z. c/ 500m² R. 1-10.382, fls. 163 L. 2-BE. Aforado lote 3 q. 39 - 4ª z. urbana. c/ 500 m² R. 1-10.861, fls. 63 2-BH. Matriculada a área 4.000m² R. 11.075 fls. 78- 2-BI. Matriculada a área de 4.452,04m² R. 11.076, fls. 79 L. 2-BI. Matriculada a área de 25.000,00m² R. 11.077, fls. 80 L. 2-BI. Matriculada a área de 5.507.625m² R. 11.078. fls. 81 L. 2-BI. Matriculada a área de 6.492.375m² R. 11.079 fls. 82. L. 2-BI. Matriculada a área de 312,50m² R. 1- 11.728, fls. 144 2-BM. Doada a área de 13.855,00m² lote 130, 1ª z. suburbana R. 1-11.902 fls. 120 2-BN. Aforado os lotes 6,7 e 8. q. 3-A, c/ 3.000m² R. 1- 12.502, fls. 130 L. 2-BQ. Aforado o lote 7 q. 44, 4ª z. c/ 1.000m² R. 1-12.563 fls. 192 L. 2-BQ. Aforado o lote 01 q. 134, c/ 1.000m² R. 1- 12.287. L. 2-BP fls. 111, a Artinda A. Pereira. Aforada o lote 6, q. 60, c/ 1.000m² R. 1-12.756, L. 2-BR. fls. 187 Salustiano G. Costa. Aforado metade lote 97, 2ª z. sub. c/ reg. 1-12.940, 2-BS fls. 172 c/ 48.400m². Aforado metade lote 5-B, Q. 125, c/ 300m² Reg. 1-13.035, 2-BT fls. 65, c/ 48.400m² 17/12/82. Aforado metade lote 5/M. q. 76, c/ 710,23m² reg. 1-13.066, 2-BT fls. 102 c/ 48.400m² 23/12. Doada p/ lote 130 c/ 12.000m² conf. R. 1-13.190, L. 2-BU. fls. 27 ao Dom Bosco Futebol Clube. Em 13/01/83. Aforado o lote n. 06 q.62 c/ 1.000m² conf. R. 1- 13.536. L. 2-BV, fls. 182, a Sebastião Mendes. Aforado os lotes 8 e 9. q. 104, 2ª z. urbana c/ 2.000m², conf. reg. 1-13.594, fls. 40 2-BX, 4/4/83. Doado a Associação Médica de Três Lagoas 2.400m² conf. reg. 1- 14.550 fls. 22, 2- CD em 01/12/83. Doado ao sindicato dos empregados E. Bancários 2.400m² conf. reg. 1-14.960 fls. 70 2-CF 8/3/84. Aforado a A.P.A.E. parte do lote sub. 130 c/ 13.855m² conf. reg. 1-15.955 fls. 62, 2-CL, 22/10/84. Aforado a Zuque & Cia 36.453,00m², conf. reg. 1-16.420, L.02, ficha 01. Aforado o lote 1. q. 107 B.N.S. Aparecida c/ 1.000m² conf. reg. 1-16.671 L. 02 ficha 1. Aforado p. lotes sub. 43 e 51 c/ 2.232m² a Walter Machado do Nascimento R.1-17.162 em 22/7/85. Destacado parte lote 2 quadra 36/1, 1ª zona c/ 500m² (12,50x40) M. 17.379, 17/9/85. Aforado p/lotes 52-51-43 c/ 2.700,00m² a Antônio Darci Ferreira R. 1-17.488 11/10/85. Aforado p/ lotes 52-51-43 c/ 1836,00m², a Jose Laurentino dos Santos, R. 1-17.489 11/10/85. Doado p/ lote 130 c/ 7.355m² a Ass. Treslagoense de Professores, R. 1-17.879 15/1/86. Aforado lote 4, qd. 119 c/ 1.000m² a Loja Maçônica Renascença 42. R- 1.18.980. Aforado lote 05

Miriam Reis Costa
Oficial Registradora Titular

Edson Reis Costa Filho   *Nilson Cilene Reis Costa*   *Miriam Cleusa Reis Costa*   *Giovani Gomes Roman*   *Silvina de Lima Valim*
1º Ofício Substituto   2º Ofício Substituto   Escrevente Autorizado   Escrevente de Registro

Av. Antonio Trajano, 1126, TRÊS LAGOAS/MS – Fone/Fax (67) 3921-3204

quart. 116 c/ 1.000m² a Miguel Donato R.1-19.020 23/9/86. Aforado o lote 03 quart. 116 c/ 1.000m² a Sebastiana M. da Conceição R.1-19.743. Aforado lote 05 quart. 33 4ª zona c/ 1.000m² Est. M. Grosso do Sul, 29-7-87. Aforado parte lote 63, 1ª zona sub. c/ 680,78m² a Dinaura Camargo Corniani conf. R.1/21.127, livro 2 ficha 01 em 07/01/88. Aforado parte lote 6 qdra. 66 1ª zona suburb. c/ 176,50m² a Alison Dias Ferreira, conf. R. 21.853, livro 2 ficha 01 em 17-6-88. Aforado parte lote 130 1ª zona urb. c/ 6.000,00m² a Associação dos Agentes trib. de Mato Grosso do Sul- conf. R. 1/21.703 livro 2 fls. 1 em 30-6-88. Aforado parte lote 02 (dois) da quadra 03 na 2ª zona urb. c/ 350,00m² a Donato Francisco dos Santos conf. R.01/23.435 livro 2 fls. 01 em 10/8/89. Aforado parte lote 05 quart. João Carrato, c/ 170,00m² (10x17ms) a Selma Gehre, conf. R. 01/26.079, livro 02, fls. 01 em 26/4/91. Aforado lote 12 quadra 01 5ª zona c/ 400,00m² (10x40) a João Desidério dos Santos em 23/06/92. Matriculada a área de 1.906.35m², conforme matricula n. 31.308, livro 02, fls. 01 em 20-04-95. Parte do lote 88 1ª zona suburbana. Aforada parte lote 01, quadra 33, 4ª zona urb. a Osvaldo Tosta de Queiroz conf. R. 01/ M.33.198, em 26/08/97. Matriculada ao Município parte lote 02, quadra 38/01,1ª z. urb. com 700,00m2 (18,50x40) M. 34.043, em 7/4/98. Matriculada ao Município lote com a área de 23.510,25m² conf. M. 38.129, livro 02, f. 01, em 09/05/2001. Av.T.462.- Protocolo: 145.370 em 19/03/2010.- Pelo processo de Suscitação de Dúvida. Autos n. 021.10.004737-9, figurando como suscitante: Mirian Reis Costa, e como interessado: Município de Três Lagoas/MS, e de acordo com a decisão proferida pela MM. Juíza de Direito da Vara da Fazenda Pública e Registros Públicos, Dra. Aline de Oliveira Lacerda, faz-se a presente averbação para constar a abertura de matrícula para o imóvel urbano constituído pelo Lote "A", com a área de 3.676,286m², perímetro 273,53m², localizado entre as quadras 13, 14 e 15 do loteamento Jardim das Acácias e a área de propriedade de Magid Thomé, neste Município e Comarca de Três Lagoas/MS, conforme matricula n. 69.111, do livro 02, deste registro imobiliário. Eu, Giovani Gomes Roman, Escrevente autorizado, digitei e conferi. Dou fé. Três Lagoas/MS, 18 de junho de 2014. Oficial/Substituto/Escrevente Autorizado. (ass) Giovani Gomes Roman. Av.T.462.- Protocolo: 145.371 em 19/03/2010.- Pelo processo de Suscitação de Dúvida. Autos n. 021.10.004737-9, figurando como suscitante: Mirian Reis Costa, e como interessado: Município de Três Lagoas/MS, e de acordo com a decisão proferida pela MM. Juíza de Direito da Vara da Fazenda Pública e Registros Públicos, Dra. Aline de Oliveira Lacerda, faz-se a presente averbação para constar a abertura de matrícula para o imóvel urbano constituído pelo Lote "B", com a área de 3.613,165m², perímetro 272,26m², localizado entre as quadras 13, 14 e 15 do loteamento Jardim das Acácias e a área de propriedade de Magid Thomé, neste Município e Comarca de Três Lagoas/MS, conforme matricula n. 69.112, do livro 02, deste registro imobiliário. Eu, Giovani Gomes Roman, Escrevente autorizado, digitei e conferi. Dou fé. Três Lagoas/MS, 18 de junho de 2014. Oficial/Substituto/Escrevente Autorizado. (ass) Giovani Gomes Roman. Av.T.462.- Protocolo: 145.372 em 19/03/2010.- Pelo processo de Suscitação de Dúvida, Autos n. 021.10.004737-9, figurando como suscitante: Mirian Reis Costa, e como interessado: Município de Três Lagoas/MS, e de acordo com a decisão proferida pela MM. Juíza de Direito da Vara da Fazenda Pública e Registros Públicos, Dra. Aline de Oliveira Lacerda, faz-se a presente averbação para constar a abertura de matrícula para o imóvel urbano constituído pelo Lote "C", com a área de 5.305,287m², perímetro 370,74m², localizado entre as quadras 13, 14 e 15 do loteamento Jardim das Acácias e a área de propriedade de Magid Thomé, neste Município e Comarca de Três Lagoas/MS, conforme matrícula n. 69.113, do livro 02, deste registro imobiliário. Eu, Giovani Gomes Roman, Escrevente autorizado, digitei e conferi. Dou fé. Três Lagoas/MS, 18 de junho de 2014. Oficial/Substituto/Escrevente Autorizado. (ass) Giovani Gomes Roman. **Av.T.452** - Protocolo: 180.619 em 26/05/2014.- Pelo processo de Suscitação de Dúvida, Autos n. 021.10.004737-9, figurando como suscitante: Mirian Reis Costa, e como interessado: Município de Três Lagoas/MS, e de acordo com a decisão proferida pela MM. Juíza de Direito da Vara da Fazenda Pública e Registros Públicos, Dra. Aline de Oliveira Lacerda, faz-se a presente averbação para constar a abertura de matrícula para o imóvel urbano constituído pelo Lote "A", da **Quadra "41"**, com a área de 2.590,00m², destinado ao "Paço Municipal", localizado na 1ª zona urbana "Centro", neste Município e Comarca de Três Lagoas/MS, conforme matrícula n. 69.114, do livro 02, deste registro imobiliário. Eu, Giovani Gomes Roman, Escrevente autorizado, digitei e conferi. Dou fé. Três Lagoas/MS, 18 de junho de 2014. Oficial/Substituto/Escrevente Autorizado. (ass) Giovani Gomes Roman. **Este é o teor do traslado.** Em 04 de fevereiro de 2022. Eu, *Alessandra Amaro Bernardo*, auxiliar extrajudicial, digitei. Eu, **Miriam Clarice Reis Costa**, 3ª Oficial Substituta, conferi o teor. O referido é verdade e dou fé.

<u>CERTIDÃO</u>

Certifico que a presente cópia é reprodução fiel da Transcrição n. 442 Livro 3-A, fls. 112, requerida por Amanda Emiliana Santos B., inscrita no CPF/MF n. 453.773.538-85, para a finalidade de Pesquisa de Mestrado. **ADVERTÊNCIA**: A presente certidão não se presta a qualquer outro fim, especialmente ALIENAÇÃO ou GRAVAÇÃO do imóvel. Dou fé. Três Lagoas/MS, 08/02/2021

<u>CUSTAS</u>

Emolumentos	R$ 29,00
FUNJECC	R$ 2,90
FUNDE-PGE	R$ 1,16
FUNADEP	R$ 1,74
FEADM/MS	R$ 2,90
SELO	R$ 1,50
TOTAL	R$ 38,20

<u>SELO DIGITAL</u>

AGC99874-159-NO4L

Este selo poderá ser conferido e autenticado no site www.tjms.jus.br/corregedoria/selo/autenticidade.php

Oficial/Substituto/Escrevente autorizado

REPÚBLICA FEDERATIVA DO BRASIL
ESTADO DE MATO GROSSO DO SUL
SERVIÇO DE REGISTRO DE IMÓVEIS

Miriam Reis Costa
Oficial do Registro de Imóveis

f Reis Costa Filho
1 Escrivão - Designado

Miriam Cilene Reis Costa
2º Oficial Substituto

Miriam Clarice Reis Costa
3º Oficial Substituto

Giovani Gomes Roman
Escrevente Autorizado

Sabrina de Lima Valim
Escrevente de Registro

Av. Antonio Trajano, 1320, TRÊS LAGOAS/ MS – Fone/ Fax: (67) 35212247

CERTIFICO, em razão de meu cargo e a pedido verbal de parte interessada, que, revendo neste Cartório o fichário e **Livro 3-F, f. 33**, verifiquei constar o registro com o seguinte teor:

N. de Ordem: 2.385. **Data:** 28 de novembro de 1936. **Circunscrição:** Município de Três Lagoas. **Denominação ou rua e Número do Imóvel:** "Santa Helena". **Característicos e Confrontações:** Propriedade rural denominada "Santa Helena", com a área de quatro mil e cinquenta hectares (4.050) e tem a configuração de um polígono irregular, achando-se os respectivos marcos colocados: o 1º fica a margem direita da Estrada de Ferro Noroeste do Brasil; o 2º também a margem direita da mesma Estrada; deste ponto a linha divisória deixa a Estrada de Ferro e vai fazendo divisa com os lotes dos colonos Vicente e Alonso, a 8.169 metros e 30 centímetros do 1º em diferentes rumos; o 3º a 80 metros do 2º, no rumo 80°09'NO; o 4º a 200 metros do 3º, no rumo 88°19'NO; o 5º a 1.465 metros do 4º no rumo 20°24'NO; o 6º a 330 metros do 5º no rumo 69°36'NE; o 7º a 1.080 metros do 6º no rumo 20°24'SE; o 8º a 158 metros do 7º no rumo 0°42'SE; o 9º a 240 metros do 8º no rumo 64°21'SE; e fica no ponto terminal da Estrada de Ferro Noroeste do Brasil, no ponto onde começa o levantamento do rio Paraná; o 10º a 2.685 metros do 9º no rumo 41°44'NO e fica no pontal do rio Sucuriú no Paraná; o 11º a margem direita do rio Sucuriú a 4.996 metros e 40 centímetros do 10º em diferentes rumos servindo de limite entre esses dois marcos o rio Sucuriú; o 12º a 460 metros e 50 centímetros do 11º no rumo 42°20 SO; o 13º a 3.047 metros e 80 centímetros do 12º no rumo 22°46'SO e fica na divisa da fazenda Varginha; o 14º a 2.907 metros do 13º no rumo 41°38'NO; o 15º a 6.296 metros e 50 centímetros do 14º no rumo 43°10'NO; o 16º a 83 metros e 60 centímetros do 15º no rumo 74°14'SO; o 17º a 159 metros do 16º no rumo 43°40'SO; o 18º a 167 metros e 20 centímetros do 17º no rumo 4°52'SE; o 19º a 881 metros e 30 centímetros do 18º no rumo 1°3'SO; o 20º a 1.050 metros e 10 centímetros do 19º no rumo 39°7'SE; o 21º a 949 metros e 80 centímetros do 20º no rumo 6°32'SE; o 22º a 1.179 metros do 21º no rumo 8°22'SO; o 23º a 7442 metros do 22º no rumo 53°43'SE e a 3.766 metros do 1º no rumo 36°17'SO, como tudo consta do memorial e planta. Eu, Carlos Correa da Costa, sub-oficial, o escrevi. Eu, Felipe Neri Monteiro, Oficial subscrevo e assino. Três Lagoas, 28 de novembro de 1936. (a) Felipe Neri Monteiro. **Nome, Domicílio, Profissão, Estado e Residência do Adquirente:** COMPANHIA FEIRA DE GADO DE TRÊS LAGOAS, em liquidação, com sede na cidade de São Paulo. **Nome, Domicílio, Profissão, Estado e Residência do Transmitente:** O ESTADO DE MATO GROSSO. **Título de transmissão:** Título definitivo. **Forma do título, Data e Serventuário:** Título definitivo expedido pelo Secretário de Terras e Obras Públicas em 17 de novembro de 1936, e assinado pelo Governador Dr. Mario Correa da Costa e M. C. Oliveira Mello. **Valor do Contrato:** Rs. 200:000 (duzentos contos de reis). **Condições do Contrato:** As das leis sobre terras. **Averbações:** Vide transcrição n. 3.257, livro 3-H, f.2.

Vide transcrição n. 3.432, livro 3-H, f.83. Vide transcrição n. 3.001, livro 3-G, f.59.
Vide transcrição n. 3.002, livro 3-G, f.60. Vide transcrição n. 3.003, livro 3-G, f.60. Vide transcrição n. 3.018, livro 3-G, f.65.

Em tudo dai graças
1 Ts 5.18

Miriam Reis Costa
Oficial do Registro de Imóveis

Edward Reis Costa Filho
Oficial Substituto Desposto

Miriam Cilene Reis Costa
2ª Oficial Substituta

Miriam Clarice Reis Costa
3ª Oficial Substituta

Giovani Gomes Roman
Escrevente Autorizado

Sabrina de Lima Velin
Escrevente de Registro

Av. Antônio Trajano, 1329, TRÊS LAGOAS/ MS – Fone/ Fax: (67) 35212247

Vide transcrição n. 3.019, livro 3-G, f.66. Vide transcrição n. 3.075, livro 3-G, f.85. Vide transcrição n. 3.113, livro 3-G, f.96.
Vide transcrição n. 3.124, livro 3-G, f.100. Vide transcrição n. 3.142, livro 3-G, f.105. Vide transcrição n. 3.160, livro 3-G, f.112.
Vide transcrição n. 3.166, livro 3-G, f.115. Vide transcrição n. 3.180 e 3.181, livro 3-G, f.121.
Vide transcrição n. 3.215, livro 3-G, f.133. Vide transcrição n. 3.227, livro 3-G, f.138.
Vide transcrição n. 3.240, livro 3-G, f.142. Vide transcrição n. 3.593, livro 3-H, f.157. Vide transcrição n. 3.640, livro 3-H, f.182.
Vide transcrição n. 4.954, livro 3-K, f.146. Vide transcrição n. 4.842, livro 3-K, f.104. Vide transcrição n. 10.454, livro 3-U, f.195.
Vendido 7,32,25 há conf. transc. n. 11.787, f. 90, livro 3-Z.

Este é o teor do traslado. Em 04 de fevereiro de 2022. Eu, *Alessandra Amaro Bernardo*, **auxiliar extrajudicial, digitei. Eu,** *Miriam Clarice Reis Costa*, 3ª Oficial Substituta, **conferi o teor. O referido é verdade e dou fé.**

<table>
<tr><td>

CERTIDÃO

Certifico que a presente cópia é reprodução fiel da Transcrição n. 2.385 Livro 3-F, fls. 33, requerida por Amanda Emiliana Santos B., inscrita no CPF/MF n. 453.773.538-46, para a finalidade de Pesquisa de Mestrado. **ADVERTÊNCIA:** A presente certidão não se presta a qualquer outro fim, especialmente **ALIENAÇÃO** ou **GRAVAÇÃO** do imóvel. Dou fé. Três Lagoas/MS, 08/02/2021

CUSTAS

Emolumentos	R$ 29,00
FUNJECC	R$ 2,90
FUNDE-PGE	R$ 1,16
FUNADEP	R$ 1,74
FEADMP/MS	R$ 2,90
SELO	R$ 1,50
TOTAL	R$ 39,20

</td><td>

SELO DIGITAL

AGC99875-S03-IIOR

Este selo poderá ser conferido e autenticado no site: www.tjms.jus.br/corregedoria/selos/pesquisaSelo.php.

</td></tr>
</table>

Em tudo das graças
1 Ts 5,18

MATRÍCULA	FOLHA
-40.035-	**-01-**

TRES LAGOAS/ MATO GROSSO DO SUL

LIVRO Nº 2   REGISTRO GERAL

REPÚBLICA FEDERATIVA DO BRASIL
ESTADO DE MATO GROSSO DO SUL
CARTÓRIO DO 1º OFÍCIO

AV. FERNÃO MÜLLER, 1.519 - CEP 79600-002 - FONE (067) 521-2247 - FAX (067) 521-9135 - TRÊS LAGOAS - MS

Imóvel: lote suburbano sob n.01, localizado no Núcleo Agro-Industrial Vargiona, denominada Fazenda Reflorestamento e Pecuária São Thomé, neste Município e Comarca, com a área de 242,38,40ha. (duzentos e quarenta e dois hectares, trinta e oito ares, quarenta centiares), com as seguintes medidas e confrontações: inicia-se no marco 01 cravado na lateral da rodovia MS-158, no lugar onde faz divisa com o Aeroporto Municipal, e a zona urbana desta cidade, e dali segue pela cerca de arame confrontando com o Aeroporto Municipal no rumo 68°24'46"NE com 546,60 metros até o marco 02 cravado no canto da cerca. Do marco 02 prossegue a direita tendo como divisa a cerca de arame e confrontando com Antônio do Carmo nos rumos e distâncias: 46°34'00"SE – 476,66 metros; 46°22'23"NE – 255,46 metros; 44°08'26"NW – 51,90 metros; 46°45'56"NE – 502,61 metros; 62°18'02"NE – 49,79 metros; 43°15'06"SE – 242,78 metros; 43°26'00"NE – 828,83 metros, passando pelos marcos 03,04,05,06 e 07, encontra o marco 08 cravado no canto da cerca. Do marco 08 prossegue a direita tendo como divisa a cerca de arame e confrontando com o Aeroporto Municipal no rumo 66°00'00"NE com 430,00 metros até o marco 09 cravado no canto da cerca. Do marco 08 prossegue a direita tendo como divisa a cerca de arame e confrontando com o Aeroporto Municipal no rumo 66°00'00"NE com 430,00 metros até o marco 08 cravado no canto da cerca. Do marco 09 prossegue a direita tendo como divisa a cerca de arame e confrontando com a área da Prefeitura Municipal no rumo 13°33'00"SW com 2.526,80 metros até o marco 10 cravado no canto da cerca. Do marco 10 prossegue a direita tendo como divisa a cerca de arame e confrontando com Neltexsul Industria Têxtil Ltda. e Cortex Industria Têxtil Ltda. com os rumos e distâncias: 46°34'00"NW – 876,50 metros; 43°26'00"SW – 345,72 metros, passando pelo marco 11 encontra o marco 12 cravado na lateral da Rodovia MS-158, deste segue fazendo divisa com a mesma propriedade, hoje Rodovia MS – 158 com o rumo de 46°34'00"NW e a distância de 120,00 metros até o marco 13 cravado no canto da cerca. Do marco 13 prossegue à direita tendo como divisa a cerca de arame e confrontando com Murilo Tebet Thomé nos rumos e distâncias: 43°46'00"NE – 200,01 metros; 46°34'00"NW – 150,00 metros; 43°46'00"SW – 200,01 metros, passando pelos marcos 14 e 15, encontra o marco 16 cravado no canto da cerca. Do marco 16 prossegue a direita confrontando com a área da mesma propriedade hoje ocupada pela Rodovia MS 158 pela cerca de arame no rumo 46°34'00"NW e na distância de 1.163,90 metros ao marco 01, inicial e final desta descrição. Memorial descritivo elaborado em 13.09.2002, pelo engenheiro Sr. Manoel Alves de Queiroz, CREA 63468. TD/SP Visto 751 MS. Cadastro: CCIR 98/99 Código do Imóvel: 912034 012661 9. Mod Rural: 27,5ha. N.Mod Rurais: 11,15. Mod Fiscal: 35ha. N. Mod Fiscais: 8,76. F. Mín. Parc.3,0ha, área total 306,6ha. Cadastrado em nome de Magid Thome, com a denominação de Reflorestamento e Pecuária S Thome. **Registro anterior:** M.40.034, livro 02 f.01, datada de 31 de dezembro de 2002, deste Registro Imobiliário. **Proprietários:** **Dr. Magid Thomé Filho**, médico veterinário, inscrito no CPF/MF.n.306.920.821-00, casado em regime de comunhão parcial de bens, na vigência da lei 6.515/77 com a **Sra. Adriana de Castro Weiler Thomé**, de lides do lar, inscrita no CPF/MF.n.901.538.551-34, domiciliados nesta cidade, na rua Visconde de Tamandaré, n.524; **Dr. Murilo Tebet Thomé**, advogado, inscrito no CPF/MF.n.338.243.801-10, casado em regime de comunhão universal de bens, na vigência da lei 6.515/77 com a **Sra. Magali Mussa Martins Thomé**, comerciante, brasileira, inscrita no CPF/MF.n.357.529.631-68, domiciliados nesta cidade, na rua Generoso Siqueira, n.1236; **Sra. Valéria Egidio Thomé Maia**, professora, inscrita no CPF/MF.n.312.686.921-15, casada em regime de comunhão parcial de bens na vigência da lei 6.515/77 com o **Sr. Zenith Maia Vasconcellos Filho**, cirurgião dentista, inscrito no CPF/MF.n.824.763.938-15, brasileiros, domiciliados na Socrates n.853, Apt° 191, bloco 1, bairro Jardim Marachoara, em São Paulo –SP; e **Sr. Leandro Tebet Thomé**, engenheiro agrônomo, inscrito o CPF/MF.n.368.642.181-04, casado em regime de comunhão parcial de bens, na vigência da lei 6.515/77 com a **Sra. Mylene Bernardes Thomé**, administradora de empresas, inscrita no CPF/MF.n.367.471.491-49, brasileiros, domiciliados nesta cidade, na rua Generoso Siqueira, 1068. **Data:** 31 de dezembro de 2002. Matrícula aberta à requerimento dos proprietários datado de 24 de setembro de 2002. Apresentou os comprovantes de pagamento do ITR dos últimos cinco exercícios quitados, CCIR 98/99, memorial descritivo, ART, mapa que ficam arquivados em Cartório. Emolumentos: R$20,12 da abertura requerida, R$7,74 da tabela "J" e R$0,60 para FUNJECG. Eu, Giovani Gomes Roman, auxiliar extrajudicial, digitei e conferi. Dou fé. Oficial/Escrevente autorizada _______________.

Av.01/M.40.035.-Protocolo:-141.249 em **08/07/2009.-Divisão Amigável.-** Pela Escritura Pública de Divisão Amigável, lavrada no livro 191, página n. 161/162/163/164, em 09 de setembro de 2008, pelo 3. Serviço Notarial e Registral de Protestos desta cidade e comarca de Três Lagoas/MS, o imóvel objeto da presente matrícula foi dividido amigavelmente; ficando a *Gleba 1-A*, com a área de 37,33,41ha (trinta e sete hectares, trinta e três ares e quarenta e um centiares) e a *Gleba 1-B*, com a área de 21,91,64ha (vinte e um hectares, noventa e um ares e sessenta e quatro centiares), denominada **"FAZENDA REFLORESTAMENTO E PECUARIA SÃO THOMÉ"** ao condômino **MAGID THOMÉ FILHO** casado com **ADRIANA DE CASTRO WEILER THOMÉ**, conforme registro n. 01 das matrículas 52.997 e 52.998, respectivamente; a *Gleba 2-A*, com a área de 65,20,23ha (sessenta e cinco hectares, vinte ares e vinte e três centiares), denominada **"FAZENDA REFLORESTAMENTO E PECUARIA SÃO THOMÉ"** ao condômino **LEANDRO TEBET THOMÉ** casado com **MYLENE BERNARDES THOMÉ**, conforme registro n. 01 da matrícula 53.001; a *Gleba 3-A*, com a área de 65,20,35ha (sessenta e cinco hectares, vinte ares e trinta e cinco centiares), denominada **"FAZENDA REFLORESTAMENTO E PECUARIA SÃO THOMÉ"** a condômina **VALÉRIA EGIDIO THOMÉ MAIA** casada com **ZENITH MAIA VASCONCELLOS FILHO**, conforme

-continua no verso-

"Feliz a Nação cujo Deus é o Senhor"

190

registro n. 01 da matrícula 53.004; a *Gleba 4-A*, com a área de 49,49,48ha (quarenta e nove hectares, quarenta e nove ares e quarenta e oito centiares), denominada "FAZENDA REFLORESTAMENTO E PECUARIA SÃO THOMÉ" aos condôminos MURILO TEBET THOMÉ casado com MAGALI MUSSA MARTINS THOMÉ, conforme registro n. 01 da matrícula 53.006 e *Gleba 5*, com a área de 3,23,29ha (três hectares, vinte e três ares e vinte e nove centiares), denominada "FAZENDA REFLORESTAMENTO E PECUARIA SÃO THOMÉ" aos condôminos MAGID THOMÉ FILHO, casado com ADRIANA DE CASTRO WEILER THOMÉ; MURILO TEBET THOMÉ, casado com MAGALI MUSSA MARTINS THOMÉ; VALÉRIA EGIDIO THOMÉ MAIA, casada com ZENITH MAIA VASCONCELLOS FILHO; LEANDRO TEBET THOMÉ, casado com MYLENE BERNARDES THOMÉ, conforme matrícula 53.008; todas livro 02, f. 01 deste registro imobiliário, ficando portanto encerrada a presente matrícula. Eu, Jacqueline Yamaguti Ueda, Auxiliar Extrajudicial, digitei e conferi. Dou fé. Três Lagoas-MS, 05 de agosto de 2009. Oficial/Substituto/Escrevente Autorizado

Av.02/M.40.035.-Averbação Ex-Officio.-Faz-se a presente averbação para constar que por um lapso quando da averbação n. 01 desta matrícula, deixou de constar o valor dos emolumentos, sendo Emol: R$ 30,00; FUNJECC 10%: R$ 3,00; FUNJECC 3%: R$ 0,90, permanecendo encerrada a presente matrícula. Eu, Jacqueline Yamaguti Ueda, auxiliar extrajudicial, digitei e conferi. Dou fé. Três Lagoas/MS, 21 de janeiro de 2010. Oficial/Substituto/Escrevente Autorizado. -

____
Oficial do Registro / Substituto / Escrevente
Acesse o site http://www.tjms.jus.br/corregedoria/selos/pesquisaselos.php para visualizar a

MATRÍCULA **58.575** FOLHA **01**

REPÚBLICA FEDERATIVA DO BRASIL
ESTADO DE MATO GROSSO DO SUL
SERVIÇO DE REGISTRO DE IMÓVEIS

TRÊS LAGOAS / MATO GROSSO DO SUL
LIVRO Nº 2 — REGISTRO GERAL

Imóvel: Rural denominado **"FAZENDA VÓ RUTHY"**, com a área de 441,7487ha. (quatrocentos e quarenta e um hectares, setenta e quatro ares e oitenta e sete centiares) situada neste município e comarca de Três Lagoas/MS, dentro da seguinte descrição: Inicia-se a descrição deste perímetro no vértice M01; Tipo de divisa cerca; deste, segue confrontando com o Corredor Público e Rua Quexeramobim, com o azimute 120°59'34" e distância de 1.558,66m até o vértice M02; Tipo de divisa cerca; deste, segue confrontando com o Corredor Público, com os seguintes azimutes e distâncias: 210°39'01" e 411,61 m até o vértice M03; 235°34'04" e 6,38 m até o vértice M04; 259°15'09" e 6,03 m até o vértice M05; Tipo de divisa cerca; deste, segue confrontando com a faixa de domínio da ferrovia, com os seguintes azimutes e distâncias: 263°38'11" e 74,48 m até o vértice M06; 259°42'20" e 25,00 m até o vértice M07; 254°37'34" e 24,48 m até o vértice M08; 252°48'46" e 24,78 m até o vértice M09; 251°07'39" e 41,65 m até o vértice M10; 227°19'14" e 125,29 m até o vértice M11; 219°55'18" e 322,97 m até o vértice M12; 218°10'54" e 183,30 m até o vértice M13; 219°21'50" e 387,21 m até o vértice M14; 218°18'08" e 145,50 m até o vértice M15; 213°50'11" e 1.814,37 m até o vértice M16; Tipo de divisa cerca; deste, segue confrontando com a área pertencente a José Carlos de Souza Prata Tibéry Neto e Orestes Prata Tibéry Neto, matrícula n. 34.038, com os seguintes azimutes e distâncias: 309°35'19" e 296,77 m até o vértice M17; 315°01'11" e 571,57 m até o vértice M18; Tipo de divisa cerca; deste, segue confrontando com a área pertencente à Abatel - Abatedouro de Bovinos Três Lagoas Ltda., matrícula n. 26.503, com os seguintes azimutes e distâncias: 47°23'47" e 30,09 m até o vértice M19; 314°40'48" e 226,84 m até o vértice M20; 314°38'41" e 195,88 m até o vértice M21; Tipo de divisa cerca; deste, segue confrontando com a área pertencente ao Espólio de Gastone Sartori, denominada Fazenda Rodeio, matrícula n. 35.581, com o azimute 16°15'57" e distância de 764,77 m até o vértice M22; Tipo de divisa cerca; deste, segue confrontando com a área constante na matrícula n. 24.563, pertencente ao Orestes Prata Tibéry Júnior, com os seguintes azimutes e distâncias: 30°54'12" e 296,63 m até o vértice M23; 121°01'01" e 306,35 m até o vértice M24; 30°53'04" e 807,84 m até o vértice M25; Tipo de divisa cerca; deste, segue confrontando com o Corredor Público, com o azimute 30°53'09" e distância de 1.361,42m até o vértice M01, ponto inicial da descrição deste perímetro. Certidão negativa de débitos relativos ao imposto sobre propriedade territorial rural emitida em 21/06/2011 pela Secretaria da Receita Federal do Brasil, com validade até 18/12/2011, código de controle da certidão: 8BD0.6499.FB28.182E, onde consta o NIRF 2.333.381-2, Estância São José, Município de Três Lagoas/MS., área total: 212,9ha, em nome de Orestes Prata Tibery Junior. CCIR 2006/2007/2008/2009, onde consta o código do imóvel 000.051.171.646-2, área total 212,9000ha, mód.rural 40,0473, n. mód.rurais 4,22; mód.fiscal 35,0000; n. mód.fiscais 6,0800; FMP 2,0000; área registrada 212,9000ha; classificação fundiária: média propriedade produtiva; Fazenda Vó Ruth, Município de Três Lagoas/MS, em nome de Orestes Prata Tibery Junior, nacionalidade brasileira, código da pessoa 00.488.294-6. Certidão negativa de débitos relativos ao imposto sobre propriedade territorial rural emitida em 21/06/2011 pela Secretaria da Receita Federal do Brasil, com validade até 18/12/2011, código de controle da certidão: F7E7.317A.7DCC.FEA1, onde consta o NIRF 2.486.889-2, Fazenda Vó Ruth, Município de Três Lagoas/MS., área total: 277,1ha, em nome de Orestes Prata Tibery Junior. CCIR 2006/2007/2008/2009, onde consta o código do imóvel 912.034.014.397-1, área total 277,1000ha, mód.rural 40,0000, n. mód.rurais 5,50; mód.fiscal 35,0000; n. mód.fiscais 7,9100; FMP 2,0000; área registrada 277,1000ha; classificação fundiária: média propriedade produtiva; Fazenda Vó Ruth, Município de Três Lagoas/MS, em nome de Orestes Prata Tibery Junior, nacionalidade brasileira, código da pessoa 00.488.294-6. Memorial descritivo, elaborado e assinado pelo Responsável Técnico Alcindo Dias Tavares, CREA 359/D-MS, guia de ART 11284057. **Registro anterior**: Matrículas n. 22.694 e 36.513, ambas do livro 02, deste registro imobiliário. **Proprietário**: **ORESTES PRATA TIBERY JUNIOR**, RG 6.629-SSP/MT, CPF/MF 008.024.681-87, brasileiro, pecuarista, separado judicialmente, residente e domiciliado na Avenida Antonio Trajano, n. 439, Centro, nesta cidade de Três Lagoas/MS. Emolumentos: R$ 18,00; FUNJECC 10%: R$ 1,80; FUNJECC 3%: R$ 0,54. Eu, Fernanda Andrade Moura, Auxiliar Extrajudicial, digitei e conferi. Dou fé. Três Lagoas/MS, 22 de junho de 2011. Oficial/Substituto/Escrevente Autorizado. - ____________

"Feliz a Nação cujo Deus é o Senhor"

Av.01/M.58.575.- De acordo com a averbação n. 02 (dois) da matrícula n. 22.694, livro 02 deste registro imobiliário, existe no imóvel objeto da presente matrícula, 20% (vinte por cento) de Reserva Legal, em cumprimento ao que determina a Lei n. 4.771 de 15/09/65 com as alterações da Lei 7.803 de 18/07/89, não podendo dela ser feita qualquer tipo de exploração, a não ser mediante autorização do IBAMA. Eu, Fernanda Andrade Moura, Auxiliar Extrajudicial, digitei e conferi. Dou fé. Três Lagoas/MS,22 de junho de 2011. Oficial/Substituto/Escrevente Autorizado. -

AV.02/M.58.575.- De acordo com o registro n. 04 (quatro) da matrícula n. 38.513, deste Registro imobiliário, existe no imóvel objeto da presente matrícula, **HIPOTECA DE PRIMEIRO GRAU,** em favor do Banco do Brasil S/A, agência desta cidade, para garantia da dívida de R$ 468.800,00 (quatrocentos e sessenta e oito mil e oitocentos reais), pelo Aditivo de Retificação e Ratificação à Cédula Rural Hipotecária n. 20/90.310-3, datado de 31/03/2003, com vencimento para 01/03/2013. Eu, Fernanda Andrade Moura, Auxiliar Extrajudicial, digitei e conferi. Dou fé. Três Lagoas/MS,22 de junho de 2011. Oficial/Substituto/Escrevente Autorizado. -

Av.03/M.58.575. - Protocolo:- 164.175 em 27/05/2011. - **Alteração de estado civil.-** Pelo requerimento datado de 27 de maio de 2011, o proprietário **Orestes Prata Tibery Junior,** requer a presente averbação para constar a alteração de seu estado civil de **separado judicialmente** para **divorciado,** conforme faz prova cópia autenticada da certidão de casamento Termo n. 450, Livro B-15, f. 049, expedida em 04/04/2000 pelo Cartório do 2º Ofício, desta Comarca de Três Lagoas/MS, onde consta a averbação de conversão de separação judicial em divórcio consensual, de acordo com sentença proferida pelo MM. Juiz de Direito da 1ª Vara Cível desta Comarca de Três Lagoas/MS, em Substituição legal, Dr. Eduardo Magrinelli Junior, expedida nos Autos n. 97.5010004-2 e que transitou em julgado em 24/04/1997, bem como requer a presente averbação para constar a alteração de seu estado civil de **divorciado** para **casado** em virtude de haver contraído matrimônio em 19/02/2010, pelo regime de separação obrigatória de bens, de acordo com o artigo 1641, parágrafo único, item II, do Código Civil, com **Ellen Perboni Martins,** e que por consequência a mesma passou a assinar **ELLEN PERBONI MARTINS PRATA TIBERY,** conforme faz prova cópia autenticada da certidão de casamento Termo n. 12379, Livro B-82, f. 90, expedida em 19/02/2010 pelo Cartório do 2º Ofício, desta Comarca de Três Lagoas/MS. Emolumentos: R$ 17,00; FUNJECC 10%: R$ 1,70; FUNJECC 3%: R$ 0,51. Selo digital n. AAY 62678-707. Eu, Fernanda Andrade Moura, Auxiliar Extrajudicial, digitei e conferi. Dou fé. Três Lagoas/MS,24 de junho de 2011. Oficial/Substituto/Escrevente autorizado. -

Av.04/M.58.575.- Protocolo:- 166.563 em 23/08/2011.- **Termo de anuência.-** Pelo termo de anuência datado de 08 de julho de 2011, expedido pelo Banco do Brasil S.A. agência 0208-9, desta cidade de Três Lagoas/MS, inscrita no CNPJ 00.000.000/0208-93, representado por seu gerente geral, Sr. Osmar Camilo Sanches – CPF 437.226.791-68, na qualidade de "anuente credor hipotecário", **anuiu a transferência/desmembramento da área de 27.6298ha, livre de ônus,** permanecendo a área remanescente do imóvel integra, sem qualquer mácula ou alteração, a hipoteca que já onera o imóvel objeto do negócio pactuado, inclusive e principalmente quanto a sua preferência sobre outros gravames que vierem a ser posteriormente registrado a matrícula desta imóvel e, também, quanto à data de constituição original da hipoteca dada a favor do Banco do Brasil S.A. Emolumentos: R$ 34,00; FUNJECC 10%: R$ 3,70; FUNJECC 3%: R$ 1,02. Selo digital n. ABN 07260-227. Eu, Fernanda Andrade Moura, Auxiliar Extrajudicial, digitei e conferi. Dou fé, Três Lagoas/MS, 13 de setembro de 2011. Oficial/Substituto/Escrevente Autorizado. -

Av.05/M.58.575.-Protocolo:- 166.563 em 23/08/2011.- **Desmembramento.-** Pelo requerimento datado de 16 de agosto de 2011, o proprietário **Orestes Prata Tibery Junior,**

já qualificado, requer o desmembramento do imóvel objeto desta matrícula, formando a área: área° de 23,3545ha (vinte e três hectares, trinta e cinco ares e quarenta e cinco centiares), conforme matrícula 59.242; do livro 02, deste Registro Imobiliário, em virtude do imóvel ter sido descaracterizado de imóvel rural para urbano e prometido a venda. Emolumentos: R$ 34,00; FUNJECC 10%: R$ 3,70; FUNJECC 3%: R$ 1,02. Selo digital n. ABN 07261-689. Eu, Fernanda Andrade Moura, Auxiliar Extrajudicial, digitei e conferi. Dou fé. Três Lagoas/MS, 13 de setembro de 2011. Oficial/Substituto/Escrevente Autorizado. -

Av.06/M.58.575.-Protocolo:-156.553 em 23/08/2011.-Área remanescente.-Pelo requerimento datado de 16 de agosto de 2011, firmado pelo proprietário Orestes Prata Tibery Junior, já qualificado, faz-se a presente averbação para constar a descrição da área remanescente do imóvel objeto da presente matrícula de 418,3942ha, (quatrocentos e dezoito hectares, trinta e nove ares e quarenta e dois centiares), dentro da seguinte descrição: Inicia-se a descrição deste perímetro no vértice AGH-M-2843; Tipo de divisa cerca; deste segue confrontando com o Corredor Público, com o azimute 120°59'34" e distância de 556,68m até o vértice M01; Tipo de divisa cerca; deste, segue confrontando com a área remanescente pertencente a Orestes Prata Tibery Júnior, com os seguintes azimutes e distâncias: 176°53'55" e 39,78m até o vértice M06; 171°17'33" e 291,84m até o vértice M05; 210°27'34" e 20,20m até o vértice M04; 121°01'51" e 742,67m até o vértice M03; 30°28'21" e 278,19m até o vértice M02; Tipo de divisa cerca; deste segue confrontando com a Rua Quexeramobim, com o azimute 120°59'34" e distância de 52,94m até o vértice AGH-M-2844; Tipo de divisa cerca; deste segue confrontando com o Corredor Público, com os seguintes azimutes e distâncias: 210°39'01" e 411,61m até o vértice AGH-M-2845; 235°34'04" e 6,38m até o vértice AGH-M-2846; 259°15'09" e 6,03m até o vértice AGH-M-8122; Tipo de divisa cerca; deste, segue confrontando com a faixa de domínio da Ferrovia, com os seguintes azimutes e distâncias: 263°38'11" e 74,48m até o vértice AGH-M-2847; 259°42'20" e 26,00m até o vértice AGHM-2848; 254°37'34" e 24,48m até o vértice AGH-M-8084; 252°48'46" e 24,78m até o vértice AGH-M-8085; 251°0'39" e 41,65m até o vértice AGH-M-8124; 218°10'54" e 183,30m até o vértice AGH-M-8125; 219°21'50" e 387,21m até o vértice AGH-M-8126; 218°18'08" e 145,50m até o vértice AGH-M-8127; 213°50'11" e 1.814,37m até o vértice AGH-M-8128; Tipo de divisa cerca; deste, segue confrontando com a área pertencente a José Carlos de Souza Prata Tibery Neto e Orestes Prata Tibery Neto, matrícula n. 34.038, com os seguintes azimutes e distâncias: 308°35'18" e 295,77m até o vértice AGH-M-8134; 315°01'11" e 571,57m até o vértice AGH-M-8129; Tipo de divisa cerca; deste, segue confrontando com a área pertencente a Abatel - Abatedouro de Bovinos Três Lagoas Ltda., matrícula n. 26.503, com os seguintes azimutes e distâncias: 47°23'47" e 30,09m até o vértice AGH-M-8130; 314°40'48" e 226,84m até o vértice AGH-M-8131; 314°38'41" e 195,88m até o vértice AGH-M-8133; Tipo de divisa cerca; deste, segue confrontando com a área pertencente ao Espólio de Gastone Sartori, denominada Fazenda Rodeio, matrícula n. 35.581, com os seguintes azimutes e distâncias: 16°15'57" e 764,77m até o vértice M07; Tipo de divisa cerca; deste, segue confrontando com a área constante na matrícula n. 24.563, pertencente ao Orestes Prata Tibery Junior, com os seguintes azimutes e distâncias: 30°54'12" e 296,63m até o vértice M08; 121°01'01" e 306,35 até o vértice M09; 30°53'04" e 807,84m até o vértice AGH-M-8138; Tipo de divisa cerca; deste, segue confrontando com o Corredor Público, com o azimute 30°53'09" e distância de 1.361,42m até o vértice AGH-M-2843, ponto inicial da descrição deste perímetro. Memorial descritivo datado de 28 de junho de 2011, elaborado e assinado pelo responsável técnico, Alcindo Dias Tavares, CREA 39/D-MS, ART n. 11305184. Emolumentos: R$ 34,00; FUNJECC 10%: R$ 3,70; FUNJECC 3%: R$ 1,02. Selo digital n. ABN 07262-019. Eu, Fernanda Andrade Moura, Auxiliar Extrajudicial, digitei e conferi. Dou fé. Três Lagoas/MS, 13 de setembro de 2011. Oficial/Substituto/Escrevente Autorizado. -

Av.07/M.58.575.- "Ex-Officio".- Faz-se a presente averbação para constar que por um lapso quando da unificação das matrículas 647, 648, 649, 4.221, 5.544, 32.645 e 32.646, resultando a matrícula 38.513, bem como a unificação da matrícula 22.694 com a matrícula 38.513, resultando a matrícula 58.575, e ainda o desmembramento da matrícula 58.575, resultando a matrícula 59.242, deixou de constar que de acordo com o registro n. 03 (três) das matrículas n. 647, 648, 649 e 4.221, todas do livro 02, deste Registro Imobiliário, existe registrado no imóvel objeto da presente matrícula, **HIPOTECA DE PRIMEIRO GRAU**, em favor do Banco do Brasil S/A, agência desta cidade, para garantia da dívida de Cr$ 346.960,00 (trezentos e quarenta e seis mil novecentos e sessenta cruzeiros), pela cédula rural hipotecária n. EPI 78/00.304-2, datada de 01/11/1978, com vencimento para 25/10/1983. Eu, Fernanda Andrade Moura, auxiliar extrajudicial, digitei e conferi. Dou fé. Três Lagoas/MS, 26 de outubro de 2011. Oficial/Substituto/Escrevente Autorizado. - ___________

Av.08/M.58.575.-Protocolo:-157.163 de 16/09/2011.-**Cancelamento**.-Pela autorização para cancelamento de registro datada de 24 de outubro de 2011, o **BANCO DO BRASIL S/A**, agência desta cidade de Três Lagoas/MS, por seus representantes: Elvis Rodrigues Sitta - gerente de negócios, mat. 2.877.724-7 e Edmilson M. de Souza - gerente de contas mat. 2.670.808-6, **nos autorizou o cancelamento dos registros ns. 03 (três) das matrículas 647, 648, 649 e 4.221 e consequentemente a averbação n. 07 (sete) desta matrícula**. Emol. R$ 34,00; FUNJECC 10% R$ 3,40; FUNJECC 3% R$ 1,02. Selo digital n. ABS 03768-159. Eu, Fernanda Andrade Moura, auxiliar extrajudicial, digitei e conferi. Dou fé. Três Lagoas/MS, 26 de outubro de 2011. Oficial/Substituto/Escrevente Autorizado. - ___________

Av.09/M.58.575.- Protocolo:-157.662 em 07/10/2011.- **Retificação**.- Pelo requerimento datado de 03 de novembro de 2011, o proprietário **Orestes Prata Tibery Júnior**, requer a presente averbação, para constar que o número correto do seu RG é **524.096-SSP/MS** e, não como constou na abertura desta matrícula, conforme faz prova, cópia autenticada do referido documento. Emolumentos: R$ 17,00; FUNJECC 10% R$ 1,70; FUNJECC 3% R$ 0,51. Selo digital n. ABS 04503-118. Eu, Douglas Rodrigo Damasceno Fernandes, auxiliar extrajudicial, digitei e conferi. Dou fé. Três Lagoas/MS, 04 de novembro de 2011. Oficial/Substituto/Escrevente Autorizado.- ___________

Av.10/M.58.575.-Protocolo:- 157.663 em 07/10/2011.- **Desmembramento**.- Pelo requerimento datado de 16 de agosto de 2011, o proprietário **Orestes Prata Tibery Junior**, já qualificado, requer o desmembramento do imóvel objeto desta matrícula, formando a área: área de 2,2756ha (dois hectares, vinte e sete ares e cinqüenta e seis centiares), conforme **matrícula 59.656**, do livro 02, deste registro imobiliário. Emolumentos: R$ 34,00; FUNJECC 10%: R$ 3,70; FUNJECC 3%: R$ 1,02. Selo digital n. ABS 04507-636. Eu, Douglas Rodrigo Damasceno Fernandes, auxiliar extrajudicial, digitei e conferi. Dou fé. Três Lagoas/MS, 04 de novembro de 2011. Oficial/Substituto/Escrevente Autorizado.- ___________

Av.11/M.58.575.-Protocolo:-157.663 em 07/10/2011.-**Área remanescente**.-Pelo requerimento datado de 16 de agosto de 2011, firmado pelo proprietário **Orestes Prata Tibery Junior**, já qualificado, faz-se a presente averbação para constar a descrição da **área remanescente do imóvel objeto da presente matrícula de 416.1186ha, (quatrocentos e dezesseis hectares, onze ares e oitenta e seis centiares)**, dentro da seguinte descrição: Inicia-se a descrição deste perímetro no vértice AGH-M-2843; Tipo de divisa cerca; deste segue confrontando com o Corredor Público, com o azimute 120°59'34" e distância de 556,68m até o vértice M01; Tipo de divisa cerca; deste, segue confrontando com a área pertencente a Orestes Prata Tibery Júnior, com os seguintes azimutes e distâncias: 176°53'55" e 39,78m até o vértice M06; 171°17'33" e 291,84m até o vértice M05; 210°27'34" o

20,20m até o vértice M04; 121°01'51" e 742,67m até o vértice M03; Tipo de divisa cerca; deste segue confrontando com a área pertencente a Orestes Prata Tibery Júnior, com o azimute de 210°28'21" e distância de 176,93m até o vértice M10; Tipo de divisa cerca; deste, segue confrontando com a faixa de domínio da Ferrovia, com os seguintes azimutes e distâncias: 263°38'11" e 18,98m até o vértice AGH-M-2847; 259°42'20" e 25,00m até o vértice AGHM-2848; 254°37'34" e 24,48m até o vértice AGH-M-8084; 252°48'46" e 24,78m até o vértice AGH-M-8085; 251°07'39" e 41,65m até o vértice AGH-M-8086; 227°19'14" e 125,29m até o vértice AGH-M-8123; 219°55'18" e 322,97m até o vértice AGH-M-8124; 218°10'54" e 183,30m até o vértice AGH-M-8125; 219°21'50" e 387,21m até o vértice AGH-M-8126; 218°18'08" e 145,50m até o vértice AGH-M-8127; 213°50'11" e 1.814,37m até o vértice AGH-M-8128; Tipo de divisa cerca; deste, segue confrontando com a área pertencente a José Carlos de Souza Prata Tibery Neto e Orestes Prata Tibery Neto, matrícula n. 34.038, com os seguintes azimutes e distâncias: 306°35'18" e 295,77m até o vértice AGH-M-8134; 315°01'11" e 671,57m até o vértice AGH-M-8129; Tipo de divisa cerca; deste, segue confrontando com a área pertencente a Abatel – Abatedouro de Bovinos Três Lagoas Ltda, matrícula n. 26.503, com os seguintes azimutes e distância: 47°23'47" e 30,09m até o vértice AGH-M-8130; 314°40'48" e 226,84m até o vértice AGH-M-8131; 314°38'41"m até o vértice AGH-M-8133; Tipo de divisa cerca; deste, segue confrontando com a área pertencente ao Espólio de Gastone Sartori, denominada Fazenda Rodeio, matrícula n. 35.581, com os seguinte azimutes e distâncias; 16°15'57" e 764,77m até o vértice M07; Tipo de divisa cerca; deste, segue confrontando com a área constante na matrícula n. 24.563, pertencente a Orestes Prata Tibery Júnior, com os seguintes azimutes e distância: 30°54'12" e 296,63m até o vértice M08; 121°01'01" e 306,35m até o vértice M09; 30°53'04" e 807,84m até o vértice AGH-M-8138; Tipo de divisa cerca; deste segue confrontando com o Corredor Público, com o azimute 30°53'09" e distância de 1.361,42m até o vértice AGH-M-2843, ponto inicial da descrição deste perímetro. Memorial descritivo datado de 28 de junho de 2011, elaborado e assinado pelo responsável técnico, Alcindo Dias Tavares, CREA 359/D-MS, ART n. 11313568. Certidão negativa de débitos relativos ao imposto sobre a propriedade territorial rural, emitida em 08/08/2011, pela Secretaria da Receita Federal do Brasil, com validade até 04/02/2012, código de controle da certidão: C567.2C1E.62AF.E6BC, onde consta o NIRF 2.486.889-2, Fazenda Vo Ruth, Município de Três Lagoas/MS, área total: 510,8ha, em nome de Orestes Prata Tibery Junior. CCIR-2006/2007/2008/2009 onde consta o código do imóvel 912.034.014.397-1, área total 416,1186ha, mód.rural 40,0000, n. mód.rurais 8,25, mód.fiscal 35,0000, n. mód.fiscais 11,6891, FMP 2,0000, área registrada 441,7487ha, classificação fundiária: media propriedade produtiva, Fazenda Vo Ruth, município Três Lagoas/MS, em nome de Orestes Prata Tibery Junior, brasileiro, código da pessoa 00.489.294-6. Emolumentos: R$ 34,00; FUNJECC 10%: R$ 3,40; FUNJECC 3%: R$ 1,02. Selo digital n. ABS 04508-076. Eu, Douglas Rodrigo Damasceno Fernandes, auxiliar extrajudicial, digitei e conferi. Dou fé. Três Lagoas/MS, 04 de novembro de 2011. Oficial/Substituto/Escrevente Autorizado.-

Av.12/M.58.575.- Vendida a área de 48,4000ha, a **VIIV EMPREENDIMENTOS IMOBILIARIOS – SPE TRÊS LAGOAS LTDA**, conforme registro n. 01 da matrícula 62.714, livro 02, deste Registro Imobiliário. Eu, Simone de Lima Moreira, auxiliar extrajudicial, digitei. Eu, Douglas Rodrigo Damasceno Fernandes, auxiliar extrajudicial, conferi. Dou fé. Três Lagoas/MS, 03 de setembro de 2012. Oficial/Substituto/Escrevente Autorizado.

Av.13/M.58.575.-Protocolo:-164.468 em 07/08/2012.-Área remanescente.- Pela escritura pública de venda e compra lavrada no livro n. 764, f. 338, em 17 de abril de 2012, pelo Registro Civil das Pessoas Naturais e Tabelião de Notas, Distrito de Engenheiro Schimidt, comarca de São Jose do Rio Preto/SP; faz-se a presente averbação para constar a área remanescente de 367,7186ha (trezentos e sessenta e sete hectares setenta e um ares e oitenta e seis centiares) do imóvel objeto desta matrícula, com o seguinte roteiro e confrontação: inicia-se a descrição deste perímetro no vértice AGH-M-2843; Tipo de divisa cerca; deste, segue confrontando com o Corredor Público, com azimute 120°59'34" e distância de 333,84m até o vértice M13; Tipo de divisa de cerca; deste, segue confrontando com a Área 1, pertencente a Orestes Prata Tibery Junior, com os seguintes azimutes e distancias: 210°53'09" e 622,68m até o vértice M12; 120°53'09" e 1.060,34m até o vértice M11; Tipo de cerca; deste, segue confrontando com a faixa de domínio da Ferrovia com os seguintes azimutes e distancias: 227°19'14" e 52,80m até o vértice AGH-M-8123; 219°55'18" e 322,97m até o vértice AGH-M-8124; 218°10'54" E 183,30m até o vértice AGH-M-8128; 219°21'50" e 387,21m até o vértice AGH-M-8126; 218°18'08" e 145,50m até o vértice AGH-M-8127; 213°50'11" e 1.814,37m até o vértice AGH-M-8128; Tipo de divisa de cerca; deste, segue confrontando com a área pertencente a Jose Carlos de Souza Prata Tibery e Orestes Prata Tibery Neto, matrícula n. 34.038, com os seguintes azimutes e distancias: 308°35'18" e 295,77m até o vértice AGH-M-8134; 315°01'11" e 571,67m até o vértice AGH-M-8129; Tipo de divisa cerca; deste, segue confrontando com a área pertencente a Abatei – Abatedouro de Bovinos Três Lagoas Ltda., matrícula n. 26.503, com os seguintes azimutes e distancias: 47°23'47" e 30,09m até o vértice AGH-M-8130; 314°40'48" e 226,84m até o vértice AGH-M-8131; 314°38'41" e 195,88m até o vértice AGH-M-8133; Tipo de divisa cerca; deste, segue confrontado com a área pertencente ao Espolio de Gastoni Sartori, denominada "Fazenda Rodeio", matrícula n. 35.581, com o seguintes azimutes e distancias: 16°15'57" e 764,77m até o vértice M07; Tipo de divisa cerca; deste, segue confrontando com a área constante na matrícula n. 24.563, pertencentes a Orestes Prata Tibery Junior, com os seguintes azimutes e distancias: 30°54'12" e 296,63m até o vértice M08; 121°01'01" e 306,35m até o vértice M09; 30°53'04" e 807,64m até o vértice AGH-M-8138; Tipo de divisa cerca; deste, segue confrontando com o Corredor Publico, com o azimute de 30°53'09" e distancia de 1.361,42m até o vértice AGH-M-2843; ponto inicial da descrição deste perímetro. Memorial Descritivo, elaborado e assinado pelo Engenheiro Civil Pedro Donizete Bortolote, CREA 0600994590, ART n. 11380290. Emolumentos. R$ 34,00; FUNJECC 10% R$ 3,40; FUNJECC 3% R$ 1,02. Selo digital n. ADL 65690-274. Eu, Simone de Lima Moreira, auxiliar extrajudicial, digitei. Eu, Douglas Rodrigo Damasceno Fernandes, auxiliar extrajudicial, conferi. Dou fé. Três Lagoas/MS, 03 de setembro de 2012. Oficial/Substituto/Escrevente Autorizado.

Av.14/M.58.575.- Faz-se a presente averbação para constar que a área remanescente de 367.7185ha (trezentos e sessenta e sete hectares, setenta e um ares e oitenta e seis centiares) foi matriculada sob n. 62.715, livro 02, deste Registro Imobiliário; ficando, portanto, encerrada a presente matrícula. Eu, Simone de Lima Moreira, auxiliar extrajudicial, digitei. Eu, Douglas Rodrigo Damasceno Fernandes, auxiliar extrajudicial, conferi. Dou fé. Três Lagoas/MS, 03 de setembro de 2012. Oficial/Substituto/Escrevente Autorizado.

MATRÍCULA 63.078 — FOLHA 01

REPÚBLICA FEDERATIVA DO BRASIL
ESTADO DE MATO GROSSO DO SUL
SERVIÇO DE REGISTRO DE IMÓVEIS

TRÊS LAGOAS / MATO GROSSO DO SUL
LIVRO Nº 2 — REGISTRO GERAL

AV. ANTÔNIO TRAJANO, 1320 - CEP 79901-003 - FONE (67)3521-6791 - FAX (67)3521-2247 - TRÊS LAGOAS/MS

IMÓVEL. - Urbano destinado ao "SHOPPING NAÇÕES TRÊS LAGOAS", constituído pela "Gleba n. 01", desmembrada da "Fazenda Santa Helena", com a área de 12,00ha. (doze hectares), neste município e comarca de Três Lagoas/MS, dentro das seguintes medidas e confrontações: Começa a divisa desta gleba de terras, parte da Fazenda Santa Helena (área urbana), no marco M-00A (zero A), cravado à 35,00m do eixo da rodovia BR-158. Do marco M-00A, a divisa segue, com o azimute 314°17'55" e distância de 400,00m., alcançando o marco M-01 (um), cravado à 35,00m., do eixo da rodovia BR-158, tendo, dividido até este, com terras de Jefferson J. Salomão e Outros, (área destinada a faixa de domínio da BR – 158). Do marco M-01, a divisa segue com o azimute 46°56'55" e distância de 300,00m., alcançando o marco M-02 (dois), tendo dividido até este, com terras de Jefferson J. Salomão e Outros (matrícula 59.852). Do marco M-02, a divisa segue com o azimute 134°17'55" e distância de 400,00m., alcançando o marco M-03B (três B), tendo dividido até este, com terras de Jefferson J. Salomão e Outros, (área destinada a Prefeitura Municipal para o acesso do Shopping). Do marco M-03B, a divisa segue com o azimute de 226°56'55" e distância de 300,00m., alcançando o marco M-00A (zero A), tendo dividido até este, com terras de Jefferson J. Salomão e Outros, (área destinada ao acesso do Shopping e também alargamento da Estrada que leva ao Balneário Municipal) onde deu início esta descrição. Memorial descritivo datado de 10 de fevereiro de 2011, elaborado e assinado pelo agrimensor Antonio da Mota, CREA n. 896/D - MT VISTO n. 866 - MS. Guia ART n. 11319070. **Registro anterior**: Matrícula 63.077, livro 02, deste Registro Imobiliário. **Proprietários**: **JEFFERSON JORGE SALOMÃO**, brasileiro, separado judicialmente, pecuarista, portador da cédula de identidade RG n. 138.284-SSP/MS e do CPF/MF n. 178.578.211.87, residente e domiciliado na Av. Antonio Trajano dos Santos, n. 700, Centro, nesta cidade de Três Lagoas-MS; e **HELENA JORGE SALOMÃO NERY**, brasileira, pecuarista, portadora da cédula de identidade RG n. 249.994-SSP/MS e do CPF/MF n. 110.618.711-34, casada sob o regime de separação de bens, antes da vigência da Lei 6.515/77, com **CARLOS FERNANDO DA CÂMARA NERY**, brasileiro, médico, portador da cédula de identidade RG n. 536.908-SSP/PR e inscrito no CPF/MF sob n. 731.519.398-72, residentes e domiciliados na Avenida Antônio Trajano dos Santos, n. 680, Centro, nesta cidade de Três Lagoas/MS. Emolumentos: R$ 18,00; FUNJECC 10% R$ 1,80; FUNJECC 3% R$ 0,54. Eu, Giovani Gomes Roman, escrevente autorizado, digitei. Eu, Fernanda Andrade Moura de Almeida, auxiliar extrajudicial, conferi. Dou fé. Três Lagoas/MS, 27 de setembro de 2012. Oficial/Substituto/Escrevente Autorizado.

Av.01/M.63.078.- De acordo com a averbação n. 01 (um) da matrícula 63.077, livro 02, deste Registro Imobiliário, o imóvel objeto desta matrícula encontra-se no perímetro urbano deste município e comarca de Três Lagoas/MS. Eu, Giovani Gomes Roman, escrevente autorizado, digitei. Eu, Fernanda Andrade Moura de Almeida, auxiliar extrajudicial, conferi. Dou fé. Três Lagoas/MS, 27 de setembro de 2012. Oficial/Substituto/Escrevente Autorizado.

R.02/M.63.078. - Protocolo: - 168.488 em 21/01/2013. - **Integralização.** - Pela Escritura pública de Integralização de Capital de Bem Imóvel, lavrada no livro 86, f. 126/127, em 17 de janeiro de 2013, pelo Cartório do 4° Serviço Notarial e Registral de Títulos e Documentos, desta cidade e Comarca de Três Lagoas/MS, os proprietários: **JEFFERSON JORGE SALOMÃO**; e **HELENA JORGE SALOMÃO NERY**, casada com **CARLOS FERNANDO DA CÂMARA NERY**, já qualificados, onde consta que o imóvel objeto da presente matrícula foi transferido a título de integralização a empresa **SHOPPING F3 EMPREENDIMENTOS E PARTICIPAÇÕES LTDA.**, pessoa jurídica, inscrita no CNPJ/MF n. 14.183.259.0001-20, com sede na Avenida Lins de Vasconcelos, n. 2.816, Vila Mariana, na cidade de São Paulo/SP, representada por seus sócios: Jefferson Jorge Salomão; e Helena Jorge Salomão Nery, casada com Carlos Fernando da Câmara Nery, já qualificados, pelo valor de R$ 2.920,00 (DOIS MIL NOVECENTOS E VINTE REAIS). Consta na referida escritura isenção do pagamento do ITBI sobre 2% da avaliação do imóvel em R$ 840.000,00, conforme artigo 156,

§ 2° inciso I da constituição Federal de 1.888, da Lei 5.172/66, expedida em 27/11/2012, pela Prefeitura Municipal local. Apresentaram Certidão Negativa de Débitos Relativos às Contribuições Previdenciárias e às de Terceiros n. 001452013-21200259, emitida em 05/02/2013, pela Secretaria da Receita Federal do Brasil, com validade até 04/08/2013 em nome de Shopping F3 Empreendimentos e Participações Ltda; Certidão Conjunta Negativa de Débitos Relativos aos Tributos Federais e à Dívida Ativa da União, emitida em 05/12/2012, código de controle da certidão: 9082.D8B9.8812.913F, com validade até 03/06/2013 em nome de Shopping F3 Empreendimentos e Participações Ltda, emitida pela Secretaria da Receita Federal do Brasil; Certificado de Regularidade do FGTS - CRF, inscrição n. 14183259/0001-20, em nome de Shopping F3 Empreendimentos e Participações Ltda, certificação número: 201301120129512210557, emitida pela Caixa Econômica Federal, com validade até 10/02/2013. Certidão Negativa de Débitos expedida em 15 de janeiro de 2013, pela Prefeitura Municipal local. Imóvel cadastrado na Prefeitura sob n. 5.34.000.1200.12000. Emolumentos: R$ 914,00; FUNJECC 10% R$ 91,40; FUNJECC 3% R$ 27,42. Selo digital n. AEF 4T894-465. Eu, Lucas Vinícius de Oliveira Arruda, auxiliar extrajudicial, digitei. Eu, Douglas Rodrigo Damasceno Fernandes, auxiliar extrajudicial, conferi. Dou fé. Três Lagoas/MS, 06 de fevereiro de 2013. Oficial/Substituto/Escrevente Autorizado. -

---

R.03/M.63.078. - Protocolo: - 170.767 em 26/04/2013. - Integralização. - Pela Escritura pública de Integralização de Capital de Bem Imóvel, lavrada no livro 67, f. 55/56, em 19 de março de 2013, pelo Cartório do 4° Serviço Notarial e Registral de Títulos e Documentos, desta cidade e Comarca de Três Lagoas/MS, a proprietária: SHOPPING F3 EMPREENDIMENTOS E PARTICIPAÇÕES LTDA., já qualificada, representada por seus sócios: Jefferson Jorge Salomão; e Helena Jorge Salomão Nery, casada com Carlos Fernando da Câmara Nery, já qualificados, onde consta que o imóvel objeto da presente matrícula foi transferido a título de Integralização a empresa VÉRTICO TRÊS LAGOAS EMPREENDIMENTO IMOBILIÁRIO LTDA., inscrita no CNPJ/MF n. 13.335.046/0001-03, com sede na Avenida Chedid Jafet, n. 222, Edifício Millenium Park, conj. 41, sala 76, bloco D, Tower I, na cidade de São Paulo/SP, representada por: Dante Alberto Jemma Cobucci, brasileiro, solteiro, maior, administrador de empresas, portador do RG n. 29.631.797-4-SSP/SP, inscrito no CPF/MF n. 219.832.998-04; Nilton Bertuchi, brasileiro, casado, advogado, portador da cédula de identidade RG n. 23.292.880-0-SSP/SP, inscrito no CPF/MF n. 195.514.838-47, ambos com endereço comercial na Avenida Chedid Jafet, n. 222, Edifício Millenium Park, conj. 41, sala 76, bloco D, Tower, na cidade de São Paulo/SP. Constam na referida escritura certidão negativa da Fazenda Pública Municipal; certidão negativa de débitos relativos às contribuições previdenciárias e às de terceiros n. 001452013-21200259, expedida pela Receita Federal; certidão conjunta negativa de débitos relativos aos tributos federais e à dívida ativa da união, expedida pela Receita Federal; certificado de regularidade do FGTS-CRF, expedida pela Caixa Econômica Federal; e o pagamento do ITBI no valor de R$ 16.800,00, sobre 2% da avaliação do imóvel em R$ 840.000,00, conforme guia n. 87827820, expedida em 07/03/2013, pela Prefeitura Municipal local. Imóvel cadastrado na Prefeitura sob n. 5.34.000.1200.12000. Emolumentos: R$ 914,00; FUNJECC 10% R$ 91,40; FUNJECC 3% R$ 27,42. Selo digital n. AEY 20296-203. Eu, Lucas Vinícius de Oliveira Arruda, auxiliar extrajudicial, digitei. Eu, Glenda M. dos Santos Costa, auxiliar extrajudicial, conferi. Dou fé. Três Lagoas/MS, 21 de maio de 2013. Oficial/Substituto/Escrevente Autorizado. -

---

Av.04/M.63.078.-Protocolo:- 175.877 em 08/11/2013.- Alteração de tipo societário.- Pelo requerimento datado de 24 de outubro de 2013, assinado pelos diretores: Nilton Bertuchi, brasileiro, casado, advogado, portador da cédula de identidade RG n. 23.292.880-0-SSP/SP, inscrito no CPF/MF n. 195.514.838-47, residente e domiciliado na Rua Iguatemi, n. 192, 21° andar, conjunto 212, na cidade de São Paulo/SP. e Dante Alberto Jemma Cobucci, brasileiro,

199

solteiro, maior, administrador de empresas, portador da cédula de identidade RG n. 29.631.797-4-SSP/SP, inscrito no CPF/MF n. 219.832.998-04, residente e domiciliado na Rua Iguatemi, n. 192, 21º andar, conjunto 212, na cidade de São Paulo/SP, os quais juntaram a Ata da Assembleia de transformação da sociedade empresária, datada de 05 de fevereiro de 2013, registrado na Junta Comercial do Estado de São Paulo, sob n. 3530045733-1 e n. 378.914/13-3, e seus Anexos I e II, onde consta a transformação do tipo jurídico da sociedade de **Vértico Três Lagoas Empreendimento Imobiliário Ltda**, para **VÉRTICO TRÊS LAGOAS EMPREENDIMENTO IMOBILIÁRIO S.A**, inscrita no CNPJ/MF n. 13.335.046/0001-03, sociedade por ações, com sede na Avenida Dr. Chucri Zaidan, n. 920, 16º andar, sala 05, Market Place Tower I, Vila Cordeiro, na cidade de São Paulo/SP. Apresentou: Certidão Simplificada, emitida em 30 de outubro de 2013, pela Junta Comercial do Estado de São Paulo; Certidão Negativa de Débitos Relativos às Contribuições Previdenciárias e às de Terceiros n. 007772013-21200046, emitida em 06/08/2013, pela Secretaria da Receita Federal do Brasil, com validade até 02/02/2014 em nome de Vértico Três Lagoas Empreendimento Imobiliário Ltda; Certidão Conjunta Negativa de Débitos Relativos aos Tributos Federais e à Dívida Ativa da União, emitida em 06/08/2013, código de controle da certidão: 5798.25F7.BC17.D955, com validade até 02/02/2014 em nome de Vértico Três Lagoas Empreendimento Imobiliário Ltda; Certificado de Regularidade do FGTS - CRF, inscrição n. 13335046/0001-03, em nome de Vértico Três Lagoas Empreendimento Imobiliário Ltda, certificação número: 2013112017055028754673, emitida pela Caixa Econômica Federal, com validade de 20/11/2013 à 19/12/2013. Emolumentos: R$ 34,00; FUNJECC 10%: R$ 3,40; FUNJECC 3%: R$ 1,02. Selo digital n. AGH 88462-277 (este selo poderá ser conferido e autenticado no site: www.tjms.jus.br/corregedoria/selos/pesquisaSelo.php). Eu, Simone de Lima Moreira, auxiliar extrajudicial, digitei. Eu, Fernanda Andrade Moura de Almeida, auxiliar extrajudicial, conferi. Dou fé. Três Lagoas/MS, 06 de dezembro de 2013. Oficial/Substituto/Escrevente Autorizado. -

R.05/M.63.078.-Protocolo:- 175.878 em 08/11/2013. - **Venda e compra**. - Pela escritura pública de venda e compra lavrada no livro 246, f. 017/018, em 24 de outubro de 2013, pelo 3º Serviço Notarial e de Protestos desta cidade e comarca de Três Lagoas/MS, a proprietária **VÉRTICO TRÊS LAGOAS EMPREENDIMENTO IMOBILIÁRIO S.A**, já qualificada; **vendeu a parte ideal correspondente a 2,32% (dois vírgula trinta e dois por cento)** do imóvel objeto da presente matrícula a **WTORRE HOTÉIS HOLDING S.A**, inscrita no CNPJ/MF n. 12.423.754/0001-33, sociedade por ações, com sede na Avenida Dr. Chucri Zaidan, n. 920, 16º andar, sala 36, Market Place Tower I, Vila Cordeiro, na cidade de São Paulo/SP; pelo preço de R$ 340.720,00 **(TREZENTOS E QUARENTA MIL, SETECENTOS E VINTE REAIS)**. Constam na escritura Certidões Negativas das Fazendas Públicas: Municipal e Federal, Certificado de Regularidade do FGTS, expedida pela Caixa Econômica Federal, Certidão Negativa de Débitos Relativos às Contribuições Previdenciárias e as de Terceiros, Certidão Simplificada da Junta Comercial do Estado de São Paulo, bem como o pagamento do ITBI no valor de R$ 6.814,40 sobre 2% da avaliação da parte ideal correspondente a 2,32% do imóvel em R$ 340.720,00, conforme guia n. 88902160, expedida em 09/10/2013, pela Prefeitura Municipal local. *Consta na escritura que de acordo com o § 2º do art. 555 do Código de Normas da Corregedoria Geral de Justiça de Mato Grosso do Sul a outorgada compradora declara que dispensa a vendedora da apresentação de Certidão Tributária, respondendo nos termos da Lei, pelo pagamento dos débitos fiscais existentes.* Imóvel cadastrado na Prefeitura sob n. 5.34.000.1200.12000. Emolumentos: R$ 2.481,00; FUNJECC 10% R$ 248,10; FUNJECC 3% R$ 74,43. Selo digital n. AGH 88463-621 (este selo poderá ser conferido e autenticado no site: www.tjms.jus.br/corregedoria/selos/pesquisaSelo.php). Eu, Simone de Lima Moreira, auxiliar extrajudicial, digitei. Eu, Fernanda Andrade Moura de Almeida, auxiliar extrajudicial, conferi. Dou fé. Três Lagoas/MS, 06 de dezembro de 2013. Oficial/Substituto/Escrevente Autorizado. -

---

CONTINUA NO VERSO

"Feliz a Nação cujo Deus é o Senhor"

Av.06/M.63.078.- Prenotação:- 201.237 em 23/12/2016.-Pelo requerimento, datado de 20 de dezembro de 2016, a interessada **Premoldados Protendit Ltda**, empresa inscrita no CNPJ/MF n. 58.566.373/0001-04, com sede na rua José Guide, n. 341, Distrito Industrial, na cidade de São José do Rio Preto/SP, requer a presente averbação para declarar a existência de **ação de execução de título extrajudicial**, sobre a parte ideal do imóvel objeto da presente matrícula pertencente a **Vértico Três Lagoas Empreendimentos Imobiliários S.A**, ação esta que tramita na 5ª Vara Cível da Comarca de São José do Rio Preto/SP, processo n. 1058500-07.2016.8.26.0576, onde figura como exequente **Premoldados Protendit Ltda** e como executada **Shopping Três Lagoas S.A. e outros**, tendo sido atribuído o valor da causa de R$ 7.749.271,89 (sete milhões, setecentos e quarenta e nove mil, duzentos e setenta e um reais e oitenta e nove centavos). Emolumentos: R$ 44,00; FUNJECC 10% R$ 4,40; FUNJECC 5% R$ 2,20; FUNADEP 6% R$ 2,64; FUNDE-PGE 4% R$ 1,76 e FEADMP/MS 10% R$ 4,40. Selo digital n. ANC18869-347 (este selo poderá ser conferido e autenticado no site: http://www.tjms.jus.br/corregedoria/selos/pesquisaSelo.php). Eu, Melina de Andrade Nola da Cunha, estagiária, digitei, auxiliar extrajudicial, digitei. Eu, Sabrina de Lima Valim, escrevente de registro, conferi. Dou fé. Três Lagoas/MS, 28 de dezembro de 2016. Oficial/Substituto/Escrevente Autorizado. - *Mixiani Reii Rosita*

Av.07/M.63.078. Prenotação: 206.970 em 25/10/2017. **Penhora**. Pelo termo de penhora, extraído dos autos n. 1058500-07.2016.8.26.0576, ação de execução de título extrajudicial, em que **Premoldados Protendit Ltda** move contra **Shopping Três Lagoas S.A e outros**, expedido pela 5ª Vara Cível da Comarca de São José do Rio Preto - SP, de ordem do MM. Juiz de Direito da respectiva Vara, Dr. Lincoln Augusto Casconi, **faz-se a presente averbação nós termos do Art. 844 do CPC, para constar a penhora sob 87,68% do imóvel objeto da presente matrícula**, para garantia do débito no valor de R$ 7.749.271,89 (sete milhões, setecentos e quarenta e nove mil, duzentos e setenta e um reais e oitenta e nove centavos), ficando como depositários Dante Alberto Jemma Cobucci, portador do CPF/MF n. 219.832.998-04 e Nilton Bertuchi, portador do CPF/MF n. 195.514.838-47, responsáveis legais da firma Vértico Três Lagoas Empreendimentos Imobiliários S/A, atual denominação do executado Shopping Três Lagoas S/A. Emolumentos: R$ 44,00; FUNJECC 10% R$ 4,40; FUNJECC 5% R$ 2,20; FUNADEP 6% R$ 2,64; FUNDE-PGE 4% R$ 1,76; FEADMP/MS 10% R$ 4,40. Selo digital n. AOW54482-250 (este selo poderá ser conferido e autenticado no site: www.tjms.jus.br/corregedoria/selos/pesquisaSelo.php). Eu, Sabrina de Lima Valim, escrevente de registro, digitei e conferi. Dou fé. Três Lagoas/MS, 07 de novembro de 2017. Oficial/Substituto/Escrevente Autorizado. *Mixiani Reii Rosita*

Av.08/M.63.078. Prenotação: 211.288 em 28/06/2018. **Cancelamento de Penhora** Pelo mandado de cancelamento de averbação de penhora, extraído dos autos da ação de execução de título extrajudicial n. 1058500-07.2016.8.26.0576, em que **Luiz Humberto Pereira** move contra **Shopping Três Lagoas S.A e outros**, expedida pela 5ª Vara Cível da Comarca de São José do Rio Preto/SP, proferida pelo MM. Juiz de Direito da Respectiva Vara, o Dr. Lincoln Augusto Casconi, **procedo o cancelamento de averbação de penhora sob n. 07 (sete) nesta matrícula**. Emolumentos: R$ 44,00; FUNJECC 10% R$ 4,40; FUNJECC 5% R$ 2,20; FUNADEP 6% R$ 2,64; FUNDE-PGE 4% R$ 1,76 e FEADMP/MS 10% R$ 4,40; SELO R$ 1,50. Selo digital n. AAK23900-240-NOR (este selo poderá ser conferido e autenticado no site: http://www.tjms.jus.br/corregedoria/selos/pesquisaSelo.php). Eu, Melina de Andrade Nóla da Cunha, auxiliar extrajudicial, digitei e conferi. Dou fé. Três Lagoas/MS, 11 de julho de 2018. Oficial/Substituto/Escrevente Autorizado.

R.09/M.63.078. Prenotação: 216.599 em 25/04/2019. **Venda e compra**. Pela escritura pública de venda e compra lavrada no livro 5.114, fls. 311/315, em 16 de abril de 2019, pelo 13. Tabelião de Notas, do Distrito de Campo Belo, Comarca de São Paulo/SP, a proprietária **WTORRE HOTÉIS HOLDING LTDA**, inscrita no CNPJ/MF n. 12.423.754/0001-33, com sede

—————Continua na folha 03...———

REPÚBLICA FEDERATIVA DO BRASIL
ESTADO DE MATO GROSSO DO SUL
SERVIÇO DE REGISTRO DE IMÓVEIS

Miriam Reis Costa
Oficial do Registro de Imóveis

FONE:(67)3521-4291 - FAX: (67)3521-2247
AV. ANTONIO TRAJANO, 1320 - CEP 79601-003 - TRÊS LAGOAS/MS

e foro na Avenida Presidente Juscelino Kubitschek, n. 2041, Complexo JK, Torre D, 24. Andar, sala 32, Vila Nova Conceição, na cidade e Comarca de São Paulo/SP, representada por seus procuradores **Renato Muscari Lobo**, brasileiro, casado, advogado, portador da cédula de identidade RG n. 29.294.400-7-SSP/SP, inscrito no CPF/MF n. 296.103.458-24; e **Luis Fernando Casari Davantel**, brasileiro, casado, engenheiro, portador da cédula de identidade RG n. 12.730.471-X-SSP/SP, inscrito no CPF/MF n. 085.502.588-30, ambos com endereço Comercial na Avenida Presidente Juscelino Kubitschek, n. 2041, Complexo JK, Torre D, 24. Andar, sala 32, Vila Nova Conceição, na cidade e Comarca de São Paulo/SP; **vendeu** a *parte ideal correspondente a 2,32% (dois vírgula trinta e dois por cento)*, do imóvel objeto da presente matrícula à **SHOPPING TRÊS LAGOAS LTDA**, inscrita no CNPJ/MF n. 13.335.046/0001-03, com sede na Rodovia BR 158, Km 01, s/n, esquina com a Avenida Jamil Jorge Salomão, Bairro Jardim Novo Aeroporto, nesta cidade de Três Lagoas/MS, representada por seu procurador **Pedro Henrique Carlos Vale**, brasileiro, solteiro, advogado, portador da cédula de identidade RG n. 94938627-SSP/PR, inscrito no CPF/MF n. 739.944.131-68, residente e domiciliado na Rua Pássaros e Flores, n. 223, ap. 274, Jardim das Acácias, na cidade de São Paulo/SP, pelo preço de R$ 420.000,00 (QUATROCENTOS E VINTE MIL REAIS). Constam na escritura Certidão Negativa da Fazenda Pública Federal, bem como o pagamento do ITBI, no valor de R$ 8.400,00, sobre 2% da avaliação do imóvel em R$ 420.000,00, conforme guias ns. 2898/2016 GRTM n. 124665662 e PR.5672/2019, expedidas pela Prefeitura Municipal local. *A compradora dispensa a apresentação das certidões da vendedora de que trata a Lei Federal n. 7.433/85 regulamentada pelo Decreto n. 93.240/86, com as alterações introduzidas pela Lei n. 13.097/15, inclusive de Distribuição e Feitos da Justiça do Trabalho e a Certidão Negativa de Débitos Trabalhistas, dispensando também a apresentação da Certidão Negativa de Débitos da Prefeitura Municipal, responsabilizando-se pelo pagamento de eventuais débitos em atraso.* Imóvel cadastrado na Prefeitura sob n. 7.78.000.0000.00084 – BIC 55975. Emolumentos: R$ 3.180,00; FUNJECC 10% R$ 318,00; FUNJECC 5% R$ 159,00; FUNADEP 6% R$ 190,80; FUNDE-PGE 4% R$ 127,20; FEADMP/MS 10% R$ 318,00. Selo digital n. AAB32055-824-CVD (este selo poderá ser conferida e autenticado no site: www.tjms.jus.br/corregedoria/selos/pesquisaselo.php). Eu, Simone de Lima Moreira, auxiliar extrajudicial, conferi. Dou fé. Três Lagoas/MS, 24 de maio de 2019. Oficial/Substituto/Escrevente Autorizado. -

Av.10/M.63.078. Prenotação: 219.856 em 10/10/2019. **Alteração de razão social.** Pelo requerimento datado de 04 de novembro de 2019, assinado pelo procurador **Jorge Abdul Ahad**, o qual juntou a o Contrato Social da Sociedade, datado de 29 de julho de 2016, registrado na Junta Comercial do Estado de Mato Grosso do Sul, em 12 de agosto de 2016, sob n. 54201214525, e 4ª Alteração do Contrato Social datado de 27 de junho de 2019, registrado na Junta Comercial do Estado de Mato Grosso do Sul, em 04 de outubro de 2019, sob n. 54611790, onde consta alteração de razão social de **VÉRTICO TRÊS LAGOAS EMPREENDIMENTO IMOBILIÁRIO LTDA** para **SHOPPING TRÊS LAGOAS LTDA**, inscrita no CNPJ/MF sob n. 13.335.046/0001-03. Apresentou Comprovante de Inscrição e de Situação Cadastral expedida pela Receita Federal do Brasil. Emolumentos: R$ 44,00; FUNJECC 10% R$ 4,40; FUNJECC 5% R$ 2,20; FUNADEP 6% R$ 2,64; FUNDE-PGE 4% R$ 1,76. FEADMP/MS 10% R$ 4,40; SELO R$ 1,50. Selo digital n. ACL21303-286-NOR (este selo poderá ser conferido e autenticado no site: www.tjms.jus.br/corregedoria/selos/pesquisaselo.php). Eu, Simone de Lima Moreira, auxiliar extrajudicial, conferi. Dou fé. Três Lagoas/MS, 06 de novembro de 2019. Oficial/Substituto/Escrevente Autorizado.-

Av.11/M.63.078. Prenotação: 219.856 em 10/10/2019. **Desmembramento.** Pelo requerimento datado de 09 de outubro de 2019, o proprietário **SHOPPING TRÊS LAGOAS LTDA**, requer a presente averbação para constar o desmembramento do imóvel objeto desta matrícula, formando as seguintes glebas: "Gleba 01A", com a área de 117.400,00m² (cento e dezessete mil e quatrocentos metros quadrados), conforme matrícula n. 87.205; e

—— Continua no verso... ——

"Feliz a Nação cujo Deus é o Senhor"

"Gleba 01B", com a área de 2.600,00m² (dois mil e seiscentos metros quadrados), conforme matrícula n. 87.208; ambas do livro 02, deste Registro Imobiliário; ficando, portanto, encerrada a presente matrícula. Eu, Simone de Lima Moreira, auxiliar extrajudicial, conferi. Dou fé. Três Lagoas/MS, 06 de novembro de 2019. Oficial/Substituto/Escrevente Autorizado.-

CERTIFICO que a presente fotocópia confere com a matrícula original de n. 63078 e que, nos termos do disposto artigo 19, § 1º da Lei 6.015/1973, tem valor de certidão. O referido é verdade e dou fé. Três Lagoas, MS, em 22 de dezembro de 2021. SELO nº AFU68877-138-NOR.

Oficial do Registro / Substituto / Escrevente
Acesse o site http://www.tjms.jus.br/corregedoria/selos/pesquisaselos.php para visualizar a

REFERENCES BIBLIOGRAPHIC

ABREU, S. Mato Grosso do Sul - contradictory aspects of public development policies: new/old practices... In: Lisandra Pereira Lamoso. (Org.). **Transportation and public policies in Mato Grosso do Sul.** 1 ed. Dourados: UFGD, 2008, v. 1, p. 117-134

ABREU, Silvana de. **Government planning**: SUDECO in Mato Grosso: context, purposes and contradictions. 2001. 323f. Thesis (Doctorate in Geography) - Faculty of Philosophy, Letters and Human Sciences, University of São Paulo. 2001.

ABREU, Silvana de. Rationalização e Ideologia: O domínio do capital no espaço mato-grossense. Terra Livre, v. 2, n. 21, p. 169-181, 2003.

ABREU, Silvana de. Região da Grande Dourados (MS): planning and (de)construction of a region. In: X Meeting of Latin American Geographers, 2005. **Proceedings...** University of São Paulo: São Paulo - SP, 2005.

ABRAMOVAY, R. **The social capital of territories: rethinking rural development. Seminar on Agrarian Reform and Sustainable Development.** Fortaleza: Digitized, p. 8, 1998.

ALENTEJANO, Paulo Roberto Raposo. The Bolsonaro government's policies for the countryside: counter-reform in high gear. **Revista da ANPEGE**, v. 16, n. 29, p. 351-390, 2020.

ALVES, Gilberto Luiz. Mato Grosso and History: 1870-1929 (Essay on the transition from the economic dominance of the commercial house to the hegemony of financial capital). **Boletim Paulista de Geografia**, n. 61, p. 5-82, 2017.

ALMEIDA, Rosemeire Aparecida de. Contradictions of agrarian reform in the bolsão/MS in times of paper empires. In: CAMACHO, Rodrigo Simão and COELHO, Fabiano, (0rgs.). **The Countryside in the FHC and Lula Governments: contributions to a multidisciplinary debate**. Curitiba: CRV, 2017.

ARRUDA, Larissa Rodrigues Vacari de. **MATO GROSSO'S POLITICAL ELITES:** trajectories, political practices and institutional changes 1930-1964. 2019. 275p. Thesis (Doctorate in Political Science) - Federal University of São Carlos, 2019.

To the Facts. **Why Ibama only collects 5% of the environmental fines it imposes**. Available at < https://www.aosfatos.org/noticias/por-que-o-ibama-arrecada-so-5-das-multas-ambientais-que-aplica/> Accessed on: January 24, 2022.

ACI Três Lagoas. **Our history**. Available at <https://www.acitreslagoas.com.br/nossa-historia> Accessed on: January 26, 2022.

Legislative Assembly. **Eduardo Rocha takes the platform to pay tribute to Magid Thomé.** Available at <https://al.ms.gov.br/Noticias/71479/eduardo-rocha-ocupa-a-tribuna-para-homenagear-magid-thome> Accessed on: January 25, 2022.

BARATELLI, Amanda Emiliana Santos. **The dynamics of the eucalyptus expansion process and the increase in land prices in the municipality of Três Lagoas**. 2019. 71p. (Course Conclusion Monograph) Federal University of Mato Grosso do Sul, Três Lagoas, 2019.

BARATELLI, Amanda Emiliana Santos; MILANI, Patrícia Patrícia Helana. Real estate speculation versus access to housing: we have to fight for land, fight for housing. **Electronic Journal of the Association of Brazilian Geographers Três Lagoas Section**, p. 72-96, 2019.

BITTAR, Marisa. Dream and reality: twenty-one years since the division of Mato Grosso. **Multitemas**, Campo Grande, (15): 93- 124, Oct. 1999.

BORGES, Maria Celma. Slaves, farmers and native peoples in Sant'Ana de Paranaíba: land and freedom in the fields of southern Mato Grosso (18th and 19th centuries). **Mundos do Trabalho**, v. 4, n. 8, p. 45-67, 2012.

Brasil de Fato. **Ricardo Salles: 13 facts that make the minister a threat to the planet's environment.** Available at <https://www.brasildefato.com.br/2021/04/21/ricardo-salles-13-fatos-que-fazem-do-ministro-ameaca-ao-meio-ambiente-do-planeta> Accessed on July 20, 2021.

BRAZIL. **Decree No. 9.760, of April 11, 2019**. Available at <http://www.planalto.gov.br/ccivil_03/_ato2019-2022/2019/decreto/D9760.htm> Accessed on July 20, 2021.

BRAZIL. **Complementary Law No. 87, of September 13, 1996**. Available at <http://www.planalto.gov.br/ccivil_03/leis/lcp/lcp87.htm> Accessed on July 20, 2021.

Central Bank of Brazil. **Citizen's Calculator.** Available at <https://www3.bcb.gov.br/CALCIDADAO/publico/exibirFormCorrecaoValores.do?method=exibirFormCorrecaoValores> Accessed on: January 20, 2022.

CALABI, Donatella, INDOVINA, Francesco. ON THE CAPITALIST USE OF TERRITORY. Translators: Liliana L. Fernandes and Moacyr Marques. São Paulo: USP, 1973 **(Mimeographed).**

Canal Rural. **Bolsonaro paralyzes INCRA and Agrarian Reform**. Available at <https://www.canalrural.com.br/noticias/bolsonaro-paralisar-incra-reforma-agraria> Accessed on July 20, 2021.

Canal Rural. **Government reopens negotiation for FUNRURAL debts**. Available at <https://www.canalrural.com.br/noticias/governo-reabre-negociacao-para-dividas-do-funrural-e-itr/> Accessed on July 20, 2021.

Canal Rural. **Government reopens negotiation for land debts**. Available at <https://www.canalrural.com.br/noticias/governo-reabre-negociacao-para-dividas-do-funrural-e-itr/> Accessed on July 20, 2021.

Canal Rural. **Temer government: MP 733 sanctioned, reducing some rural debts by up to 95%. See if you're one of the beneficiaries!** Available at <https://blogs.canalrural.com.br/danieldias/2016/09/30/governo-temer-sancionada-mp-733-que-reduz-em-ate-95-algumas-dividas-rurais-veja-se-voce-e-um-dos-beneficiados/> Accessed on July 20, 2021.

Chamber of Deputies. **Bill no. 6.299/2002.** Available at <https://www.camara.leg.br/proposicoesWeb/fichadetramitacao?idProposicao=46249> Accessed on: January 5, 2022.

Br banknotes. **Conversion from reis to reais.** <https://www.cedulasbr.com.br/index.php/2012-11-07-17-53-43/2012-11-07-18-05-59/2012-11-08-16-28-25> Accessed February 05, 2022.

CASTILHO, Alceu Luís. **Party of the land.** Contexto Publishing House, 2012.

CORRÊA, Roberto Lobato. **The urban network.** São Paulo: Ática, 1989.
Correa, roberto lobato. Constructing the concept of the medium-sized city. In: SPOSITO, M. Encarnação Beltrão. (Org.). **Cidades Médias** - Espaços em Transição. v. 1. São Paulo: Expressão Popular, 2007. p. 15-25.

CORRÊA, Valmir Batista. **Coronels and bandits in Mato Grosso**: 1889-1943. 2. ed. rev. e atual. Campo Grande, MS: Ed. UFMS, 2006.

CLEPS JR., João. Land reform in the context of neoliberal reforms and the political-institutional crisis in Brazil. **Revista OKARA**: Geografia em debate, v.12, n.2, p. 649-663, 2018.

Keeping an eye on the ruralistas. **Landowners owe almost R$ 1 trillion to the Union.** Available at <https://deolhonosruralistas.com.br/2016/12/12/proprietarios-de-terra-devem-quase-r-1-trilhao-uniao/> Accessed on July 20, 2021.

Diário Digital. **Três Lagoas' first shopping mall inaugurated.** Available at <https://www.diariodigital.com.br/geral/inaugurado-o-primeiro-shopping-de-tres-lagoas/#:~:text=Or%C3%A7ado%20em%20R%24%20100%20milh%C3%B5es,do%20Sul%20e%20S%C3%A3o%20Paulo> Accessed on January 27, 2022.

DELGADO, Guilherme da Costa. **From "financial capital in agriculture" to the agribusiness economy: cyclical changes in half a century (1965 - 2012).** Porto Alegre: Editora da UFRGS, 2012.

Eco. **Brazilian "chainsaw queen" attacks environmentalists**. Available at <https://oeco.org.br/reportagens/28294-rainha-da-motosserra-brasileira-ataca-ambientalistas/> Accessed on: February 1, 2022.

Eco. **Under Bolsonaro, Ibama's fines are the lowest in a decade**. Available at <https://oeco.org.br/noticias/sob-bolsonaro-autuacoes-do-ibama-sao-as-menores-em-uma-decada/> Accessed on: January 20, 2022.

FAORO, Raymundo. **The owners of power.** Editora Globo, 1958.

FABRINI, João Edmilson. Land tenure and concentration in the south of Mato Grosso do Sul. In: ALMEIDA, Rosemeire Aparecida de. **The agrarian question in Mato Grosso do Sul: A multidisciplinary view.** Campo Grande, MS: Ed. UFMS, 2008.

FABRINI, João E. Territory, class and social movements in the countryside. **Revista da ANPEGE,** v. 7, n. 07, p. 97-112, 2011.

FARIAS, Marisa de Fátima Lomba de. **América Rodrigues da Silva Camp: Hopes and Disillusions in the Memory of the Walkers Fighting for Land.** 1997. Dissertation (Masters in Sociology) - Faculty of Sciences and Letters, Universidade Estadual Paulista Júlio de Mesquita Filho. Araraquara. 1997.

FEDERAL, Senate. **City Statute. Guide for implementation by municipalities and citizens.** Brasília, 2001.

FERNANDES, Bernardo Mançano. Entering the territories of the Territory. In: **Campesinato e territórios em disputa.** São Paulo: Expressão Popular, 2008b, p. 273-302.

FERNANDES, Florestan. **A revolução burguesa: ensaio de interpretação sociológica.** Editora Contracorrente, 2020.

FONSECA. Silas Rafael da. **Latifúndio (im)produtivo e impasses à recriação camponesa no sudeste paranaense.** 2019. Thesis (PhD in Geography) - State University of Londrina, Londrina, 2019.

G1. **Understand the debate on Provisional Measure 867, which amends the Forest Code.** Available at <https://g1.globo.com/natureza/noticia/2019/05/29/entenda-o-debate-sobre-a-mp-867-que-altera-o-codigo-florestal.ghtml> Accessed on July 20, 2021.

G1. **Environment Minister defends passing 'the cattle' and 'changing' rules while media attention is focused on Covid-19.** Available at <https://g1.globo.com/politica/noticia/2020/05/22/ministro-do-meio-ambiente-defende-passar-a-boiada-e-mudar-regramento-e-simplificar-normas.ghtml> Accessed on July 20, 2021.

G1. **Ricardo Salles is investigated for illegal timber export scheme; understand.** Available at <https://g1.globo.com/df/distrito-federal/noticia/2021/06/23/ricardo-salles-entenda-operacao-contra-exportacao-ilegal-de-madeira-que-mira-ministro-do-meio-ambiente.ghtml> Accessed on July 20, 2021.

Impper Group. **OT Residential.** Available at <https://grupoimpper.com.br/empreendimentos/residencial-ot/> Accessed on February 05, 2022.

GONÇALVES, José Sidnei. The immobilization rate and the price of land: a discussion on financial speculation and patrimonial defense1. **Informações Econômicas, São Paulo,** v. 23, n. 5, p. 10-11, 1993.

HARVEY, David. **Social Justice and the City.** São Paulo: Hucitec, 1980.

HARVEY, David. **Marxist theory of the state**. In: HARVEY, David. **The capitalist production of space**. São Paulo: Anablume, 2005.

HESPANHOL, A. N. The role of the State in the Brazilian development process. In: Meneguette Junior, Messias (Org.). **FCT: 40 anos Perfil Científico Educacional.** Presidente Prudente: FCT/UNESP, 1999.

HESPANHOL, A. N. A expansão da agricultura moderna e a integração do Centro-Oeste brasileiro à economia nacional. **Caderno Prudentino de geografia**, v. 1, n. 22, p. 7-26, 2000.

Hespanhol, Nivaldo. The expansion of modern agriculture and the integration of the Brazilian Midwest into the national economy. **Caderno Prudentino de geografia**, v. 1, n. 22, p. 7-26, 2000.

Today in Três Lagoas. **"We have to be prepared for any position".** Available at <https://www.hojemais.com.br/tres-lagoas/noticia/politica/temos-que-estar-preparados-seja-para-que-cargo-for-diz-paulo-salomao-sobre-disputar-as-eleicoes-deste-ano> Accessed on: January 27, 2022.

JANNUZZI, Paulo de Martino. Social indicators in Brazil: concepts, data sources and applications. In: **Social indicators in Brazil: concepts, data sources and applications**. 2009. p. 141-141.

JP News. **Magid Thomé to be honored with name of Regional Hospital**. Available at < https://www.rcn67.com.br/jpnews/tres-lagoas/magid-thome-sera-homenageado-com-nome-de-hospital-regional/96311/> Accessed on January 27, 2022.

JP News. **If the streets told their own stories.** Available at <https://www.rcn67.com.br/jpnews/tres-lagoas/se-as-ruas-narrassem-as-proprias-historias/129381/> Accessed on: January 26, 2022.

JP News. **Orestes Prata Tibery Jr: A Zebu Poet**. Available at <https://www.rcn67.com.br/jpnews/tres-lagoas/orestes-prata-tibery-jr-um-poeta-zebuzeiro/34802/> Accessed on: January 28, 2022.

KUDLAVICZ, Mieceslau. **The territorialization of eucalyptus monoculture: a study of the eastern region of Mato Grosso do Sul**. 1. ed. Três Lagoas: Novas Edições Acadêmicas, 2014.

KUDLAVICZ, Mieceslau. **Agrarian dynamics and the territorialization of the pulp/paper complex in the micro-region of Três Lagoas. 2011.** Dissertation (Master's in Geography) - Federal University of Mato Grosso do Sul, Três Lagoas, 2011.

KUDLAVICZ, Mieceslau. Peasant stubbornness in the rural territory of bolsão. **Electronic Journal of the Association of Brazilian Geographers Três Lagoas Section**, p. 200-215, 2017.

LEAL, Victor Nunes. **Coronelismo, enxada e voto**: o município e o regime representativo no Brasil. Rio de Janeiro: Editora Nova Fronteira. 3 ed. 1997.

LEONARDO, L. A.; BARATELLI, A. E. S.; FERREIRA, J. E. B.; TEIXEIRA, J. C. Agricultura Capitalista no Território Rural Do Bolsão/Ms: expansão das monoculturas e recriação camponesa. **Terra Livre**, v. 2, n. 55, p. 273-308, 2021.

LEONARDO. Letícia Alves. **The agrarian question and religious heritage in the diocese of Três Lagoas (MS). 2020.** Dissertation (Master's in Geography) - Federal University of Mato Grosso do Sul, Três Lagoas, 2020.

LENHARO, Alcir. A Terra para quem não trabalha nela (Land speculation in western Brazil in the 1950s). **Revista Brasileira de História,** v. 6, n. 12, p. 47-64, 1986.

LOJKINE, Jean. **The capitalist state and the urban question.** Livraria Martins Fontes Editor, 1981.

LUIZ, Luana Fernanda. **Agrarian Question, National Land Credit Program and Developments for the Peasantry in the Micro-Region of Três Lagoas (MS).** Três Lagoas: 2020. 341 f. Dissertation (Master's in Geography) - Federal University of Mato Grosso do Sul, Três Lagoas, 2020.

MARX, Karl, 1818-1883. **Capital: Critique of Political Economy:** Book III: The Global Process of Capitalist Production. - 1 ed. São Paulo: Boitempo, 2017.

MARÉS, Carlos Frederico. **The social function of land.** Porto Alegre: SAFabris, 2003.

MARTINS, José de Souza. **Peasants and politics in Brazil.** Rio de Janeiro: Vozes, 1981.

MARTINS, José de Souza. **Caminhada no chão da noite**. São Paulo: Hucitec, 1989.

MARTINS, José de Souza. **The captivity of the earth.** 9 ed., 4th reprint. - São Paulo: Contexto, 2021.

MARTINS, José de Souza. **O poder do atraso: ensaios de sociologia da história lenta**. São Paulo: Hucitec, v. 2, 1994.

MATTOS, L. M. de. Fiscal austerity and the dismantling of public policies aimed at Brazilian family farming. **ANÁLISE, EMBRAPA**, 2017.

MELO, Danilo Souza; DE OLIVEIRA SILVA, Mariele. The agrarian question in the rural territory of Bolsão/MS: some approximations. **Revista Cerrados (Unimontes)**, v. 14, n. 1, p. 140-164, 2016.

MELO, D. S. **The contradictions of the reproduction of the latifundium and the (re)creation of the peasantry in the rural territories of Parque das Emas (GO) and Bolsão (MS).** 2021. 309 f. Thesis (Doctorate in Geography) - Federal University of Jataí, 2021.

MENDONÇA, Nadir Domingues. **The (un)construction of (un)orders: power and violence in Três Lagoas, 1915-1945.** Thesis (Doctorate in History). São Paulo: FFLCH-USP, 1991.

MISSIO, Fabricio José; RIVAS, Rozimare Marina Rodrigues. Aspects of the Economic Formation of Mato Grosso do Sul. **Estudos Econômicos (São Paulo)**, v. 49, n. 3, p. 601-632, 2019.

MATO GROSSO DO SUL STATE PUBLIC PROSECUTOR'S OFFICE. **Process of invasion of area no.** 08039173320148120021, of 2014.

MATO GROSSO DO SUL STATE PUBLIC PROSECUTOR'S OFFICE. **Administrative improbity case no.** 08027824920158120021, of 2015.

MORENO, Gislaene. **The paths of capitalist land appropriation in Mato Grosso.** Thesis (Doctorate in Geography) - University of São Paulo. São Paulo, 1994.

MORENO, Gislaene. **Land and Power in Mato Grosso.** Politics and Scheming Mechanisms (1892-1992). Cuiabá: EdUFMT/Entrelinhas, 2007.

MORAES, Antonio Carlos Robert. **Territory and history in Brazil.** Annablume, 2005.

MOTTA, Márcia. **Land rights in Brazil: the birth of conflict**: 1795-1824. São Paulo: Alameda, 2012.

MOREL, Marco. **The period of the Regencies (1831-1840).** Zahar, 2003.

NARDOQUE, Sedeval. Geographical expansion of capital and agrarian reform in Mato Grosso do Sul under the FHC and Lula governments. In: COELHO, Fabiano; CAMACHO, Rodrigo Simão. **The countryside in contemporary Brazil:** from the FHC government to the PT governments (agrarian question and land reform). Curitiba: CRV, 2017.

NARDOQUE, Sedeval. Agrarian question in the rural territory of bolsão/ms. In: XXIII National Meeting of Agrarian Geography, 2016. **Proceedings...** São Cristovão (SE), Nov. 2016.

NARDOQUE, Sedeval; ALMEIDA, Rosemeire A. de. Rural Territory of Bolsão (MS): reality and prospects. **NERA-Núcleo de Estudos, Pesquisas e Projetos de Reforma Agrária-Article DATALUTA. Presidente Prudente**, n. 85, 2015.

Novacana.com. **Mato Grosso do Sul: list of units in operation.** Available at: <https://www.novacana.com/usinas_brasil/estados/mato-grosso-do-sul>. Accessed on: Feb. 10, 2021.

OLIVEIRA, A. U. Non-agrarian reform and agrarian counter-reform in the Brazil of the Lula government. In: 13th Meeting of Latin American Geographers, 2011. **Proceedings**... Costa Rica, 2011a.

OLIVEIRA, Ariovaldo Umbelino de. **A agricultura camponesa no Brasil**. 2. ed. São Paulo, SP: Contexto, 1996. 164 p.

OLIVEIRA, Ariovaldo Umbelino de. The long march of the Brazilian peasantry: social movements, conflicts and Agrarian Reform. **Estudos avançados**, v. 15, n. 43, p. 185-206, 2001.

OLIVEIRA, Ariovaldo Umbelino. **Capitalist Mode of Production, Agriculture and Agrarian Reform**. São Paulo: FFLCH, 2007, 184p.

OLIVEIRA, Ariovaldo. Preface. In: ALMEIDA, Rosemeire Aparecida de. **The agrarian question in Mato Grosso do Sul**: a multidisciplinary vision. Editora UFMS, Três Lagoas, 2008.

OLIVEIRA, Francisco de. **There are open roads to Latin America**. São Paulo: Boitempo Editorial, n. 3, 2004.

OLIVEIRA, A. U. **A geografia e os movimentos sociais**. São Paulo: USP (typed), 2007.

OLIVEIRA, Gustavo de Lima Torres. Land regularization and the "global land rush" in Brazil. **Revista Campo-Território**, v. 11, n. 23 Jul., 2016.

PACHUKANIS, Evguiéni B. **General theory of law and Marxism**. Boitempo Editorial, 2017.

PASQUALINI, Juliana Campregher; MARTINS, Lígia Márcia. The singular-particular-universal dialectic: implications of the dialectical materialist method for psychology. **Psicologia & Sociedade**, v. 27, p. 362-371, 2015.

PERPETUA, Guilherme Marini. **The spatial mobility of capital and the workforce in pulp and paper production: a study from Três Lagoas (MS)**. 2012. 251 f. Dissertation (Master's in Geography). Federal University of Grande Dourados, Dourados.

PIRES, Murilo José de Souza, RAMOS, Pedro. **The Term Conservative Modernization**: Its Origin and Use in Brazil. Revista Econômica do Nordeste. Volume 40. Nº 03, July - September, 2009. Available at: https://www.bnb.gov.br/revista/index.php/ren/article/view/367#:~:text=Os%20pensa dores%20nacionais%2C%20utilizam%20o,tipicamente%20capitalistas%20na%20ag ropecu%C3%A1ria%20nacional. Accessed on: March 05, 2022.

POULANTZAS, Nicos. **O Estado, o poder, o socialismo**. 2. ed. Rio de Janeiro: Graal, 1985.

News Profile. **Orestinho's death turns five, but his legacy lives on**. Available at <https://www.perfilnews.com.br/morte-de-orestinho-completa-cinco-anos-mas-seu-legado-permanece-vivo/> Accessed on: January 27, 2022.

RAFFESTIN, C. **Por uma Geografi a do Poder**. Ática: São Paulo, 1993.

REIS, Flávio. **Political Groups and Oligarchic Structure in Maranhão**. - São Luís: [s.n.], 2007. 242 p.

RODRIGUES, Arlete Moysés. Housing in Brazilian cities. City: Contexto, 1988.

RCN 67. **Government and city hall sign R$75 million agreement.** Available at <https://www.rcn67.com.br/jpnews/tres-lagoas/governo-e-prefeitura-assinam-convenio-de-r-75-milhoes/152346/> Accessed February 8, 2022

SÃO MIGUEL, Angélica Estigarribia. **Changes in Land Use and Land Cover as a Result of the Expansion of Eucalyptus Cultivation in the Micro-regions of Três Lagoas/MS and Paranaíba/MS, in the Years 2000, 2008 and 2014.** 2015. 131 f. Dissertation (Master's in Geography). Postgraduate Program - Master's Degree in Geography, Federal University of Mato Grosso do Sul. Três Lagoas.

SANTOS, Milton. **Space and method.** São Paulo: Nobel, 1985. 88p.

SARAIVA, Flávia Carvalho Mendes. Patrimonialism and its reflexes in Brazilian public administration. **Revista Controle-Doutrina e Artigos**, v. 17, n. 2, p. 334-363, 2019.

SAUER, Sérgio et al. Bolsonaro government expands land grabbing with yet another provisional measure. **DataLuta Bulletin**, v. 144, p. 2-11, 2019.

Senate. **Bill No. 2663, of 2020.** Available at <https://www25.senado.leg.br/web/atividade/materias/-/materia/141957> Accessed on July 20, 2021.

SILVA, Cassia Queiroz da. **Free poor people in Sant'anna do Paranahyba** - 19th century. 2014. Dissertation (Master's in History) - Federal University of Grande Dourados (UFGD), Dourados, 2014.

SILVA, Ricardo Souza da. **Mato Grosso do Sul**: Labyrinths of memory. 2006. Dissertation (Master's in History) - Federal University of Grande Dourados (UFGD), Dourados, 2006.

SOUZA, A. O. **Mato Grosso do Sul in the context of the new paradigms of national integration and development.** Dourados, MS: Editora da UFGD, 2008.

SOUZA, Adáuto de Oliveira. State and induction of industrial activity: considerations on the theory of development poles in Mato Grosso do Sul. **Cadernos Acadêmicos Collection**, 2010.

SOUZA, Jessé. **The folly of Brazilian intelligence**. Casa da palavra, 2015.

SPOSITO, Maria Encarnação Beltrão. Medium-sized cities: city restructuring and urban restructuring. In: SPOSITO, M. Encarnação Beltrão. (Org.). **Cidades Médias Espaços em Transição**. v. 1. São Paulo: Expressão Popular, 2007. p. 233-253.

STEDILE, João Pedro; TRASPADINI, Roberta. Ruy Mauro Marini: life and work. **São Paulo: Expressão Popular**, 2005.

TEIXEIRA, Jodenir Calixto. The productive structure of the countryside and the evolution of livestock farming in the municipality of Três Lagoas-MS. In: III National Symposium on Agrarian Geography, 2005. **Proceedings...** Presidente Prudente - SP: Universidade Estadual Paulista Júlio de Mesquita Filho, 2005.

TEIXEIRA, Jodenir Calixto. The productive structure of the countryside and the evolution of livestock farming in the municipality of Três Lagoas-MS. In: III National Symposium on Agrarian Geography, 2005. **Proceedings...** Presidente Prudente - SP: Universidade Estadual Paulista Júlio de Mesquita Filho, 2005.

TEIXEIRA, Jodenir Calixto. The insertion of the state of mato grosso do sul in the modernization of brazilian agriculture. **National Symposium on Agrarian Geography**, v. 5, 2009.

TEIXEIRA, Jodenir Calixto; HESPANHOL, Antônio Nivaldo. The Center-West Region in the context of changes in the post-1960 period. **Revista Eletrônica AGB-TL**, v. 1, n. 3, p. 52-66, 2006.

TEIXEIRA, Jodenir Calixto; HESPANHOL, Antonio Nivaldo. Changes in farming in Três Lagoas in the context of Mato Grosso do Sul. **Caderno Prudentino de Geografia**, v. 1, n. 23, p. 246- 264, 2001.

TRÊS LAGOAS. Public Finance and Public Records Court. **Recisão/Resolução, nº 0000815-12.2009.8.12.0021** . Zenaide Aparecida Castilhos; Helio Morales Leal. Judge Aline Beatriz de Oliveira, 2009, 546p.

Economic value. **Bolsonaro wants to forgive rural debt; the gap is R$ 17 billion.** Available at <https://valor.globo.com/brasil/noticia/2018/12/21/bolsonaro-quer-perdoar-divida-rural-rombo-e-de-r-17-bi.ghtml> Accessed on July 20, 2021.

VELTMEYER, Henry; PETRAS, James. The rise and death of extractive capitalism. **Observatorio del Desarrollo**, México, v. 3, n. 9, p. 19-26, 2014.

SEE THE CITY OF TRÊS LAGOAS. **Report. Três Lagoas**, 2018

yes
I want morebooks!

Buy your books fast and straightforward online - at one of world's fastest growing online book stores! Environmentally sound due to Print-on-Demand technologies.

Buy your books online at
www.morebooks.shop

Kaufen Sie Ihre Bücher schnell und unkompliziert online – auf einer der am schnellsten wachsenden Buchhandelsplattformen weltweit! Dank Print-On-Demand umwelt- und ressourcenschonend produziert.

Bücher schneller online kaufen
www.morebooks.shop

Printed by Books on Demand GmbH, Norderstedt / Germany